APPLICATI™ TECHSERIES ELECTRONIC AUDIO CIRCUITS SOURCEBOOK

Volume 1

Disclaimer

The electronic circuits, software or related documentation in this book are NOT designed nor intended for use (whether free or sold) as on-line control equipment in hazardous environments requiring fail-safe performance, such as, but not limited to, in the operation of nuclear facilities, aircraft navigation or communication systems, air traffic control, direct life support machines or weapons systems in which the failure of the hardware or software could lead directly to death, personal injury, or severe physical or environmental damage ("high risk activities")

The author(s) and publisher(s) take no responsibility for damages or injuries of any kind that may arise from the use or misuse of the electronic circuits and/or software in this collection.

The author(s) and publisher(s) specifically disclaim any express or implied warranty or fitness for high risk activities. The electronic circuits, software and related documentation are without warranty of any kind. The author(s) and publisher(s) expressly disclaim all other warranties, express or implied, including, but not limited to, the implied warranties of merchantability and fitness for a particular purpose. Under no circumstances shall the author(s) and publisher(s) be liable for any incidental, special or consequential damages that result from the use or inability to use the circuits and software or related documentation, even if he has been advised of the possibility of such damages.

APPLICATI™ TECHSERIES ELECTRONIC AUDIO CIRCUITS SOURCEBOOK

Volume 1

APPLICATI™
www.applicati.com

Applicati Techseries
Electronic Audio Circuits Sourcebook
Volume 1

Published by

APPLICATI™

www.applicati.com

Disclaimer:

The electronic circuits, software or related documentation in this book are NOT designed nor intended for use (whether free or sold) as on-line control equipment in hazardous environments requiring fail-safe performance, such as, but not limited to, in the operation of nuclear facilities, aircraft navigation or communication systems, air traffic control, direct life support machines or weapons systems in which the failure of the hardware or software could lead directly to death, personal injury, or severe physical or environmental damage ("high risk activities")

The author(s) and publisher(s) take no responsibility for damages or injuries of any kind that may arise from the use or misuse of the electronic circuits in this collection.

The author(s) and publisher(s) specifically disclaim any express or implied warranty or fitness for high risk activities. The electronic circuits, software and related documentation are without warranty of any kind. The author(s) and publisher(s) expressly disclaim all other warranties, express or implied, including, but not limited to, the implied warranties of merchantability and fitness for a particular purpose. Under no circumstances shall the author(s) and publisher(s) be liable for any incidental, special or consequential damages that result from the use or inability to use the circuits and software or related documentation, even if he has been advised of the possibility of such damages.

Preface

Congratulations for having the first volume of ready-to-apply electronic audio circuits collection. With this source-book, you got the advantage of being able to design and assemble electronic audio modules fast and worry free. It is a sure way to optimize satisfaction in your hobby. If you are a professional electronic designer, it will help you beat the competition. Speed, efficiency, short development periods, error-free, user and maintenance friendly: these are the factors critical for success. This invaluable book filled with 84 practical ideas will help you beat project deadlines. Make your ideas work!

Make your creativity pay! JUST IN TIME!

informative...
practical...
professional...
versatile...

Acknowledgments

Many Thanks to...

E. Mischa (Optical Recognition)
D. Salinger (Electronics)
P. Schmidt (Cybernetics)
N. Lay (Robotics)

Introduction

This sourcebook contains 84 practical electronic circuits for audio applications. It contains circuits designed as amplifiers, preamps, modulators, demodulators, mixers, equalizers, signal and music generators, filters, and some auxiliary circuits commonly used in audio devices. You can combine several circuits into one large module to create a powerful electronic device specially designed for your exclusive project. Most circuits are in the amplifier category while some are auxiliary that can enhance or protect the audio circuits.

The transistors used in the circuits have more than one possible equivalents. The pin designations are also shown in details. This feature can help avoid unnecessary delays. The pins shown are either in the bottom view or front view of the transistor unless otherwise noted. Large transistors which cannot or not planned to be installed directly on the PCB must be installed on a heatsink. A dashed circle around a transistor means that the transistor must be heatsinked.

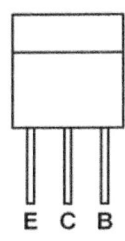

Bottom view

E C B

Front view

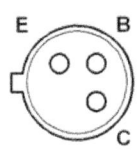

Bottom view

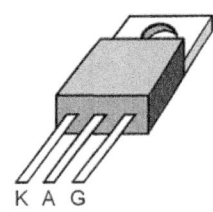

K A G

The capacitor values are given in microfarad unless otherwise specified. Electrolytic or polarized capacitors are marked with a plus sign in the diagram. This plus sign coincides with the capacitor's positive polarity in the circuit. Additionally, their voltage ratings are also given. Nonpolar capacitors are ceramic types and rated with 50 volts.

The resistor values are given in ohms (Ω), rated 1/4 watts and are of carbon film type unless otherwise specified.

The appendix pages at the end of the book contain data on semiconductor equivalents including transistors, diodes, zener diodes, FET and other multijunction semiconductors. Two pages of illustrated semiconductor pin designations and layouts are also available as appendix.

The printed circuit board layouts are also printed for the second time at the end of the book. These pages can be cut out for convenience in copying or transferring the layout on the actual pcb board.

Contents

1 AUDIO COMPRESSOR

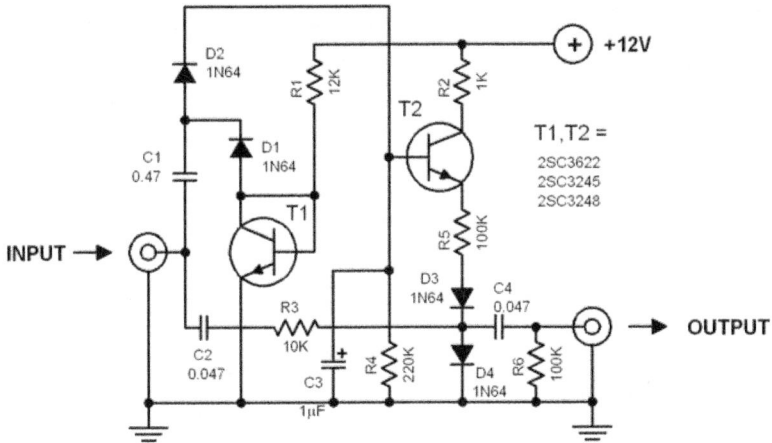

Diagram 1.0   Audio Compressor

In many audio applications, regulating the audio signal to a constant level is very important. This is true specially for voice modulated radio transmissions. Voice regulation is also often applied in intercom devices or telephone systems. Such a voice level regulator need not be very expensive.

The electronic circuit featured here combines simple design with inexpensive materials. This simple audio compressor uses a straightforward regulation technique to constantly control the amplitude of the output signal. In contrast to common feedback techniques where the output is used to regulate the input signal, the output signal of this circuit is being regulated by the input signal.

Print licensed from 123rf.com

Figure 1.0 Printed Circuit Layout

This technique simplifies the overall design of the circuit. Surprisingly, the transistor T2 works as the only active component in the circuit.

This compressor functions very well in intercom systems' or in radio transceivers. It can also be used in PA systems or telephone devices such as automatic answering machine.

Figure 1.1 Parts Placement Layout

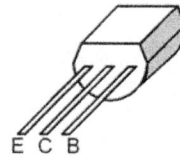

2SC3245
2SC3248
2SC3622

E C B

Transistor equivalents

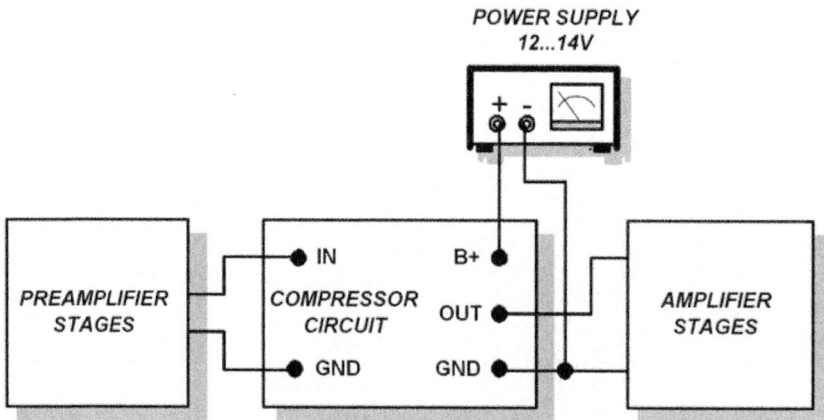

Figure 1.2 Installation Wirings

2 UNIVERSAL AUDIO AMP

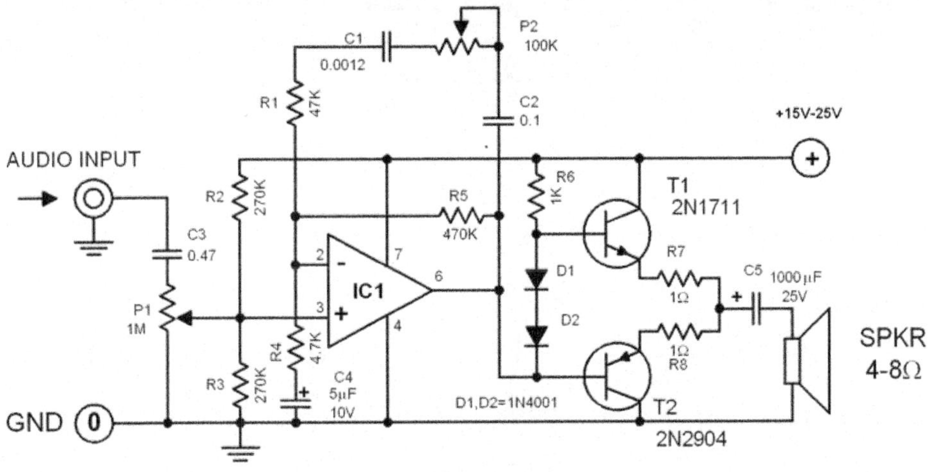

IC1= 741

Transistor equivalents:
2N2904 =2N2905
2N1711 = 2N1613,2N3901,2N3110,2N1889,2N1890

Diagram 2.0 Universal Audio Amp

Small, simple, and easy to construct. This mini-circuit is a universal audio amplifier. Just the right one for the hobbyist. It uses a standard 741 op-amp IC which is readily available from electronic supply stores. No worry about finding the components. The output power can reach slightly over 2 watts! Potentiometer P1 is the volume control while P2 controls the tone. P2 can suppress high frequencies up to maximum of 20dB. If you want to use transistors with 2S prefixes, select the transistor with specifications similar to the following:

	T1	T2
Type	npn	pnp
Collector Voltage (max)	75V	60V
Collector Current (max)	0.5A	0.6A
Power (max)	3 W	3 W
Current Gain	55	100-300
Transition Frequency(MHz)	>60	>200

Technical Data

Input Sensitivity=	100mV
Power Output=	2W
Supply Voltage=	15V - 25V
Current consumption=	150 mA

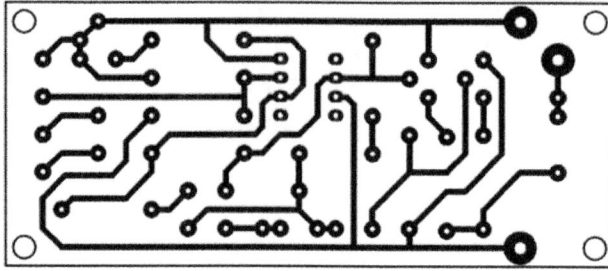

Figure 2.0 Printed Circuit Layout

Bottom view

2N1711
2N3109
2N3110
2N1889
2N1890
2N2904
2N2905

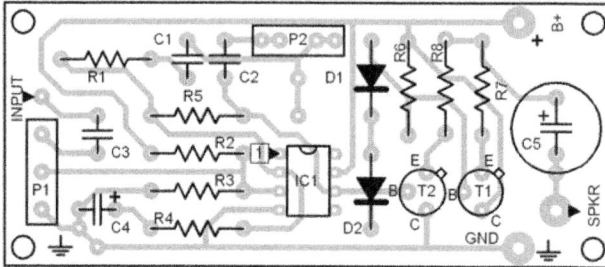

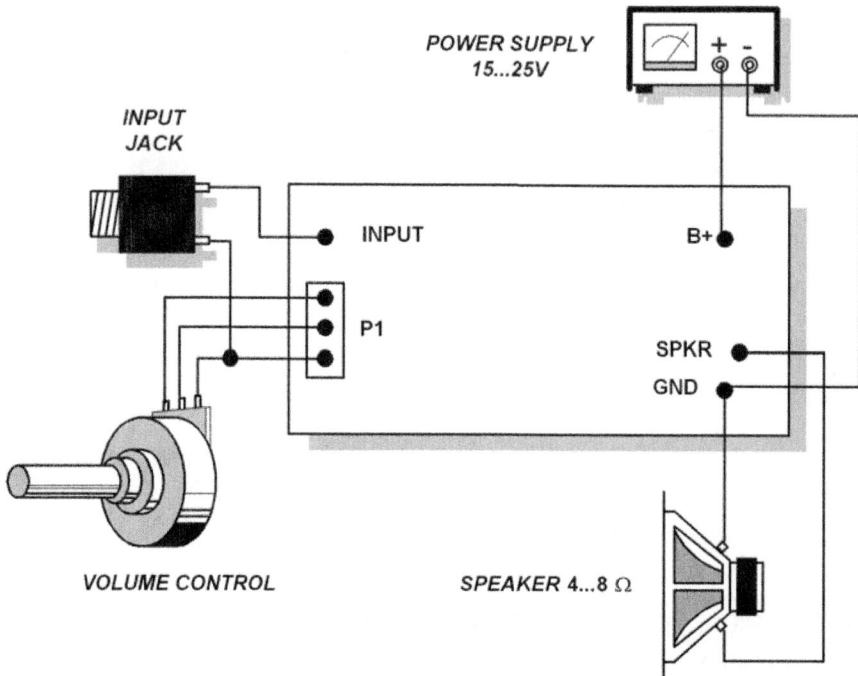

Figure 2.1 Parts Placement Layout

Figure 2.2 External Wirings

3   AUDIO MIXER

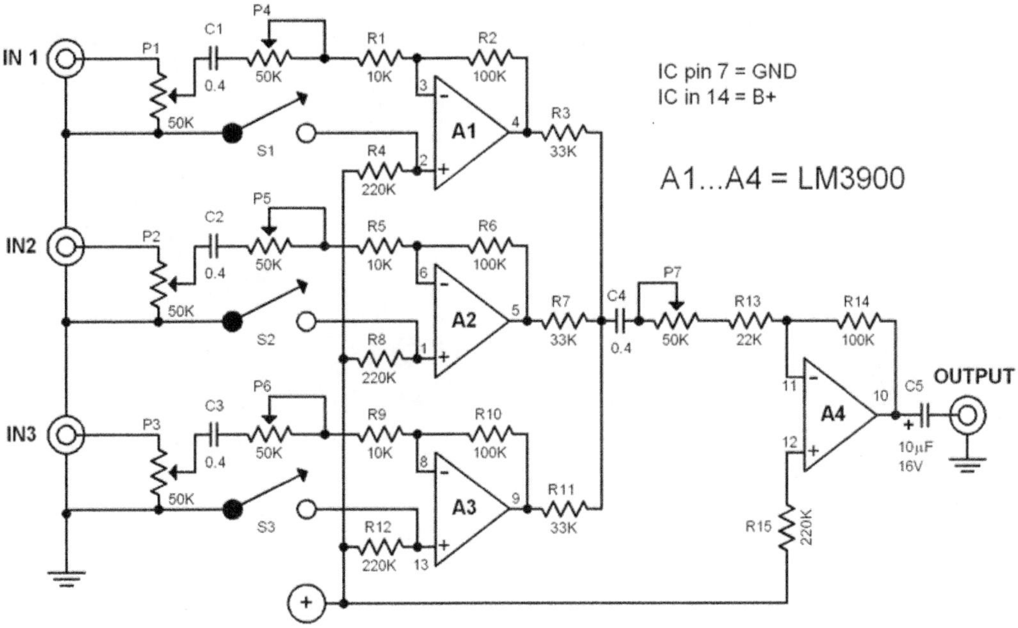

Diagram 3.0   Audio Mixer

This audio mixer circuit uses an LM3900 IC. The IC houses four integrated Norton amplifiers. The advantage of using these four op-amps is that they only need a single power supply. Since this amplifier circuit is current controlled, the DC bias is dependent on the feedback coupling. The schematic diagram shows inverting AC-Norton amplifiers. The DC output must be set at 50 percent of the power supply. In this case, a maximum output can be achieved without distortion (also called symmetrical limitation through overdrive).

In designing the circuit, you can freely choose the value of the resistor R2 (100K in the above circuit). Set the AC voltage amplification factor through the ratio of R2/R1. To set the amplifier gain correctly, choose the value of R4 = 2R2 (double the value of R2).

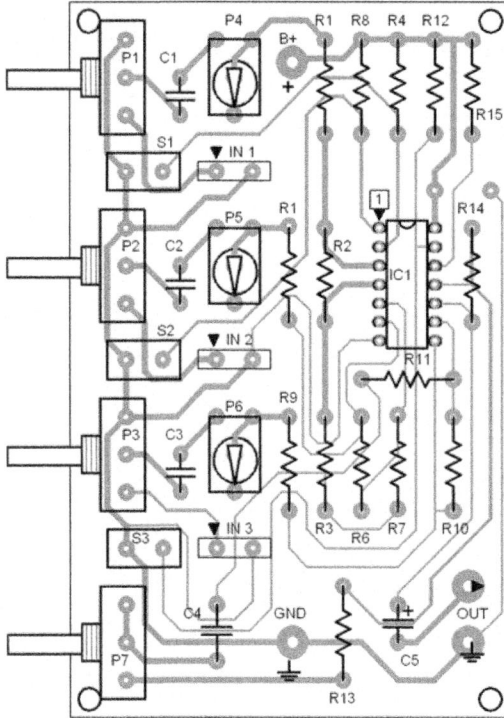

Figure 3.0 Parts Placement Layout

Figure 3.1 Printed Circuit Layout

Diagram 3.0 shows the 3-channel mixer circuit using three Norton op-amps. The input levels can be set by potentiometers P1 to P3. Furthermore, each input level can be trimmed with the help of trimmer pots P4 to P6 to adapt each input to the source. The resistors at the non-inverting inputs of the op-amps work as DC bias and set the DC output at 50 percent of the power supply. All three input signals are summed by the fourth op-amp A4 through the resistors R3, R7, and R11. The common volume level is controlled through the potentiometer P7.

+ Input	1	14 B+
+ Input	2	13 + Input
- Input	3	12 + Input
A Output	4	11 - Input
B Output	5	10 D Output
- Input	6	9 C Output
GND	7	8 - Input

LM3900
QUAD OPAMP

Figure 3.2 Pin Designations

You can switch an input channel on or off through the switches S1 to S3. An input channel is turned off when its switch is closed. It is also possible to replace these mechanical switches with transistor gates. By doing so, you can build an analog multiplexer circuit that can be easily expanded by several inputs.

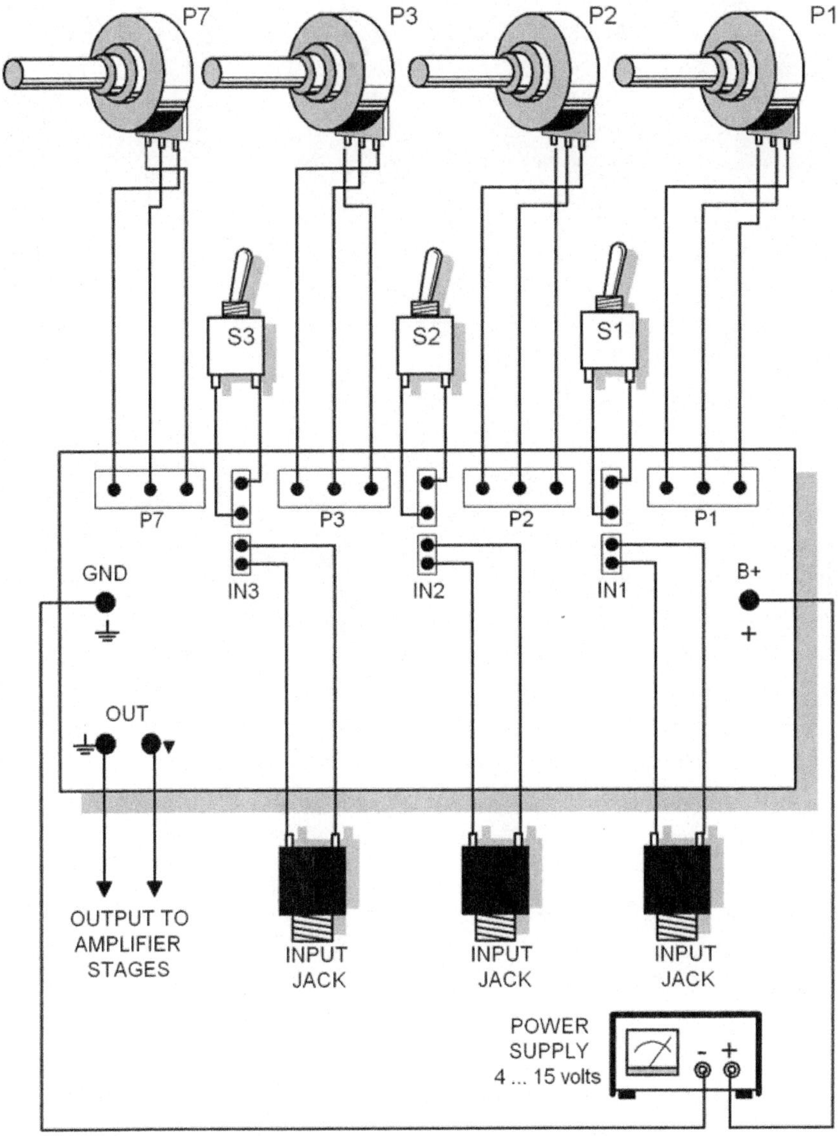

Figure 3.3 Wiring layout

Parts List:	
Resistors: R1,R5,R9 = 10K R2,R6,R10,R14 = 100K R4,R8,R12,R15 = 220K R3,R7,R11 = 33K R13 = 22K	Capacitors: C1,C2,C3,C4 = 0.4 µF C5 = 10 µF/16volts Potentiometers: P1,P2,P3,P4,P5,P6,P7 = 50K

All resistors are ¼ watts unless otherwise specified.

4 CAR STEREO AMPLIFIER

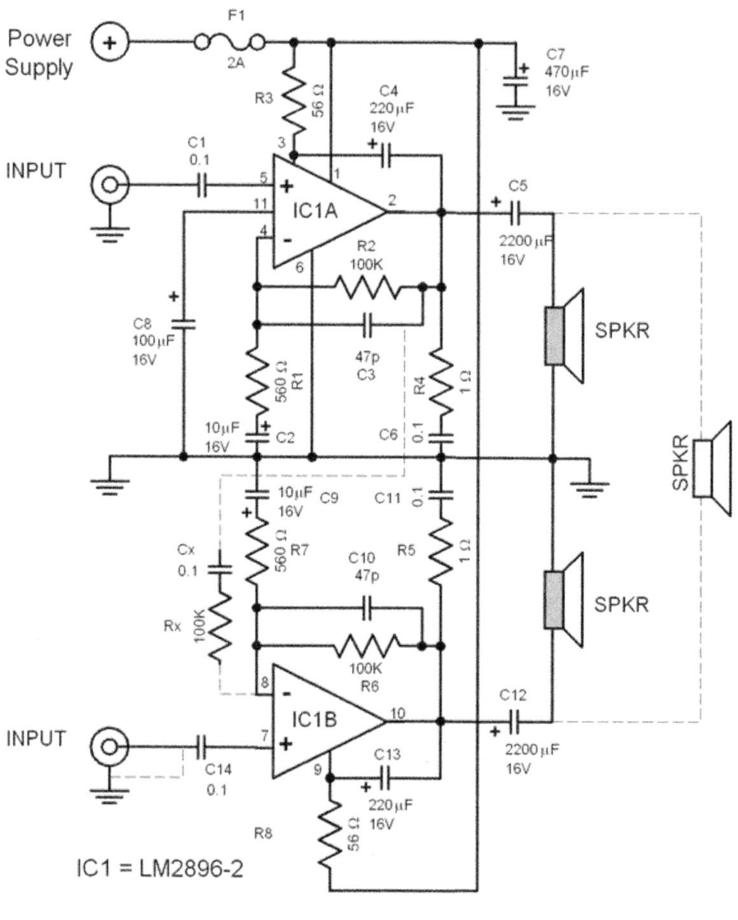

Diagram 4.0 Car Stereo Booster

This circuit uses an LM2896 IC which has two integrated amplifiers. It can be powered with voltages up to 15 volts. The power output is 2.5 watts per channel with an 8 Ω load and supply voltage of 12 volts. Using the bridge technique in the circuit produces a power output of 9 watts. The circuit can be powered with voltages from 3 volts up to 15 Volts. The load impedance that can be connected at its output can be either 4 Ω or 8 Ω. The supply voltage and the load impedance influence the output power level. See Table 4.0 for the exact data. This amplifier circuit is designed as a booster for auto radio/cassette players. The current consumption by maximum power output and a 4 Ω load is 1 ampere.

LM2896   f= 1 kHz   d=10%

POWER OUTPUT (W)

SUPPLY VOLTAGE U $_b$ (V)

RL = 4Ω BRIDGE

RL = 8Ω BRIDGE

RL = 4Ω STEREO

RL = 8Ω STEREO

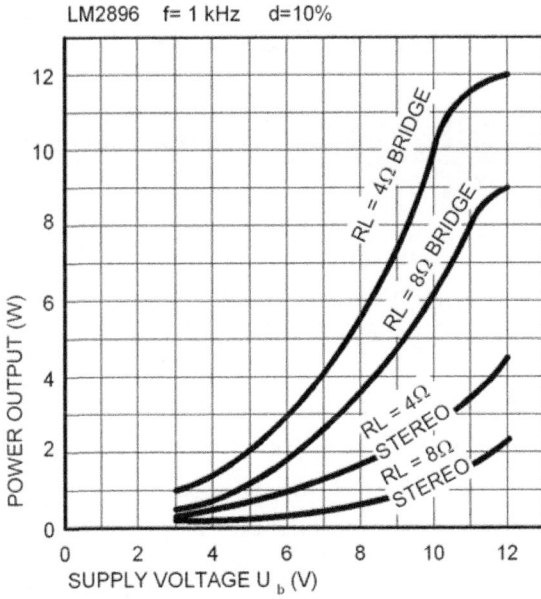

Table 4.0 Power Output vs Supply Voltage

LM 2896 -2

Pin	Designation
11	BYPASS
10	OUTPUT 1
9	BOOTSTRAP 1
8	-IN 1
7	+IN 1
6	GND
5	+IN 2
4	-IN 2
3	BOOTSTRAP 2
2	OUTPUT 2
1	+Vs

Figure 4.0 LM2896-2
Pin Designations

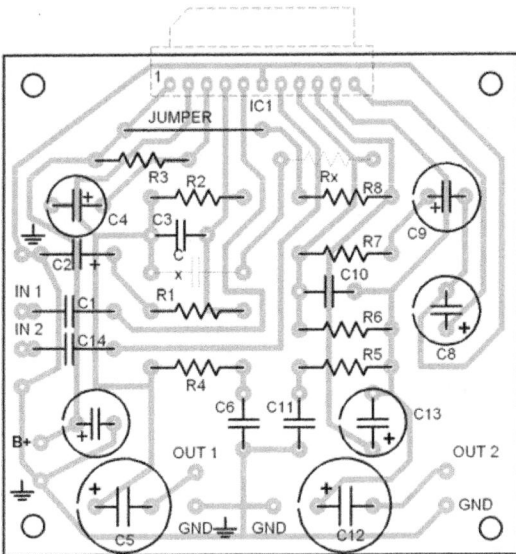

Figure 4.1 Parts Placement Layout

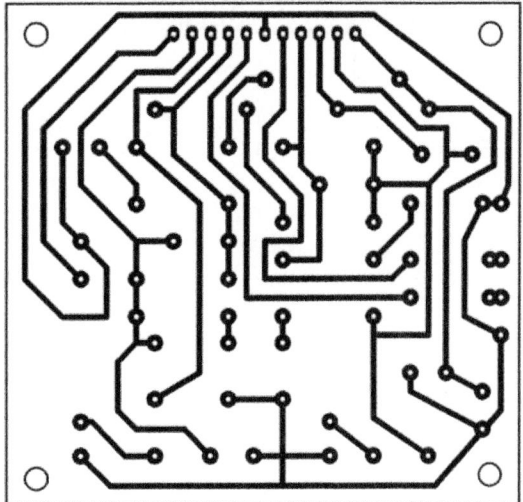

Figure 4.2 Printed Circuit Layout

To wire the booster as a bridged amplifier, follow these instructions:

Short the input number 2 of the amplifier. This is the input that is connected to the capacitor C14. See schematic diagram 4.0.

Add the additional capacitor Cx and resistor Rx to the circuit as shown by the dashed lines in the schematic diagram. These parts were considered during the design phase of the printed circuit and appropriate holes, and soldering points are already available on the circuit board.

Instead of using two speakers, use a single speaker for the amplifier. Connect the speaker to both output lines of the amplifier as shown by the dashed lines in the schematic diagram. You can use either a 4 ohm or 8 ohm speaker. The speaker impedance however affects the output power. See Table 4.0 to know what power output you will get with a certain speaker impedance.

Now, since only a single speaker is connected to the amplifier circuit, you must construct a second amplifier circuit identical to this one if you want a stereo system. The second circuit will work for the second channel (and second speaker).

Parts List:

Resistors:

R1,R7	=	560 Ω
R2,R6,Rx	=	100K
R3,R8	=	56 Ω
R4,R5	=	1 Ω

All resistors are ¼ watts unless otherwise specified.

Capacitors:
C1,C14,Cx= 0.1/50V Ceramic
C2,C9= 10 mF/16V Electrolytic
C3,C10= 47pF/50V Ceramic
C4,C13= 220 mF/16V Electrolytic
C5,C12= 2200 mF/16V Electrolytic
C7= 470 mF/16V Electrolytic
F1 = 2A Fuse

Technical Specifications:

Input sensitivity	20mV
Input impedance	100KΩ
Frequency band	
(-3dB)	30 Hz...30kHz
(-3dB) bridged	30 Hz...30kHz

Voltage amplification	180X
(bridged)	360X

Distortion factor(f=1 kHz,Ub=12V,RL=8W)
 -at 50mW output 0.095 %
 -at 1 W output 0.15 %

Power Supply	5V ... 15V
Idle Current max.	50mA
Output Power	see Table 4.0

5   AM ENVELOPE SAMPLER

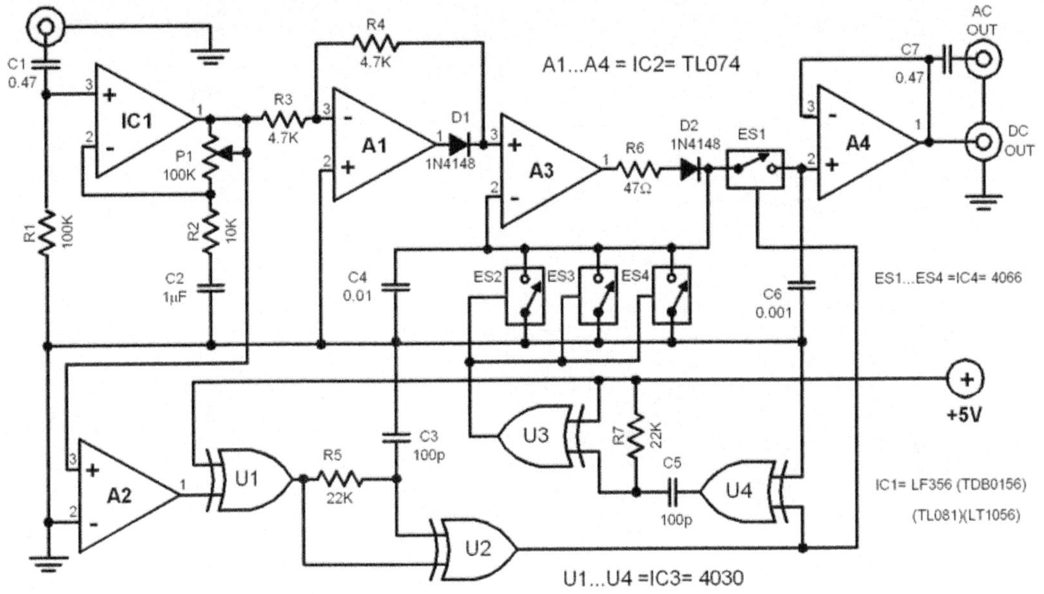

Diagram 5.0   AM Envelope Sampler

This is an AM demodulator which is not affected by phase errors in the demodulated signal. Phase error is a type of error commonly found in simple diode/low pass filter demodulators. At the input stage is an AC amplifier which is adjustable through P1. A1 and A3 are the rectifiers that charges C6 with the maximum signal voltage. The analog switches are controlled by a clock signal. The clock signal is derived from the carrier frequency. In this way the sampler can be applied in different devices like facsimile, radio and speed processors since its clock frequency is directly controlled by the received carrier frequency.

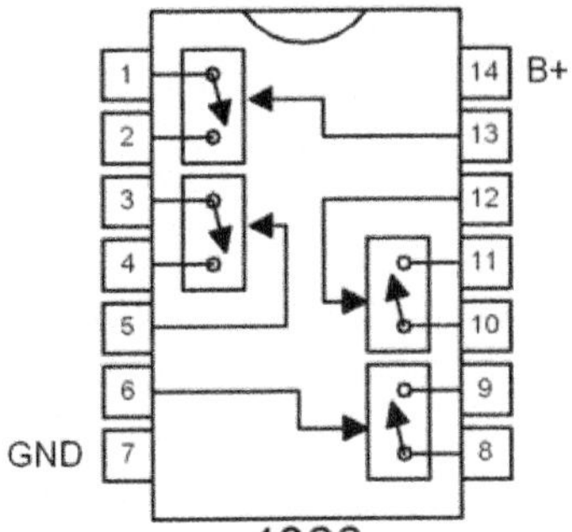

4066
Quad Digital/Analog Switch

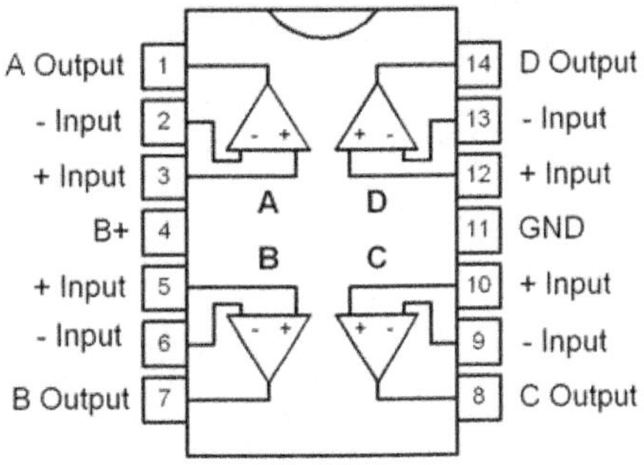

TL074
QUAD OPAMP
w/ JFET INPUT

6   CARDIOPHONE

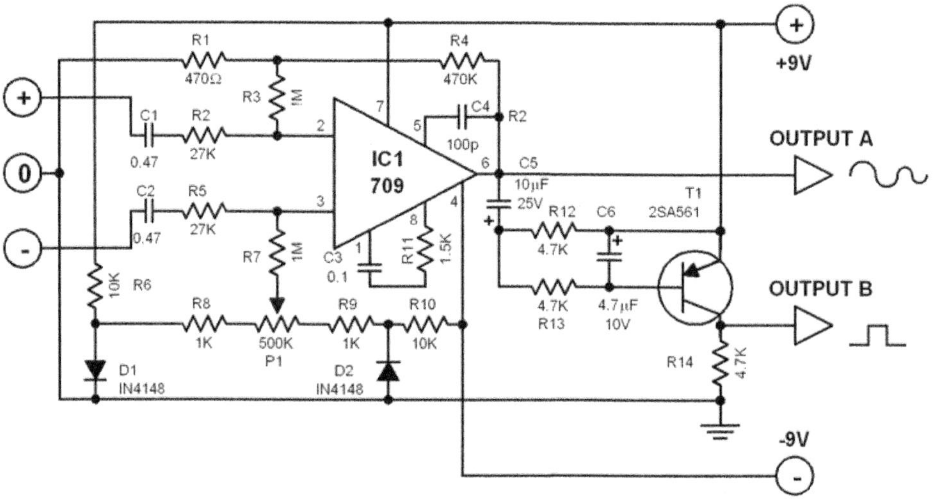

Diagram 6.0 Cardiophone

The human heartbeat can be made audible by using this cardiophone circuit! Diagram 6.0 shows the schematic of such a circuit. It is basically an audio circuit coupled to a probe made specially for the purpose of picking up the electric signal from the human heart.

To get the best signal, place the probe's electrodes to a point close to the heart. The preferred point is just below the left breast with the negative electrode pointing to the left of the sternum.

After constructing the circuit, the output A must be calibrated to null through the potentiometer P1. This is important for the circuit to function properly. The signal coming from output A can then be connected to either a low-frequency amplifier or an oscilloscope.

The signal coming from output B is a square wave in sync with the heart rhythm. This signal can be used to trigger a final amplifier or other circuits. The heartbeat can be heard from the final amplifier's speaker.

The special signal probe is shown on Figure 6.2. The simplest way to make this probe is to use a 1cm x 10 cm blank pcb board. Following the design on Figure 6.2, the non-shaded parts of the pcb board must be etched away. The un-etched copper plate must then be covered with solder to protect it from corrosion and to facilitate good contact with the skin. Take note that two of the probe's electrodes are marked negative and positive respectively. It is of utmost importance to use a shielded twisted pair wire for the cable connecting the probe with the cardiophone circuit.

Figure 6.0 Printed Circuit Layout

Figure 6.1 Parts Placement Layout

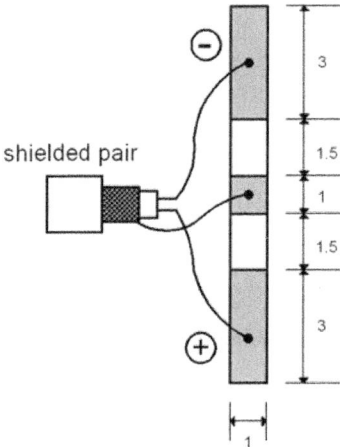

Figure 6.2 Special probe

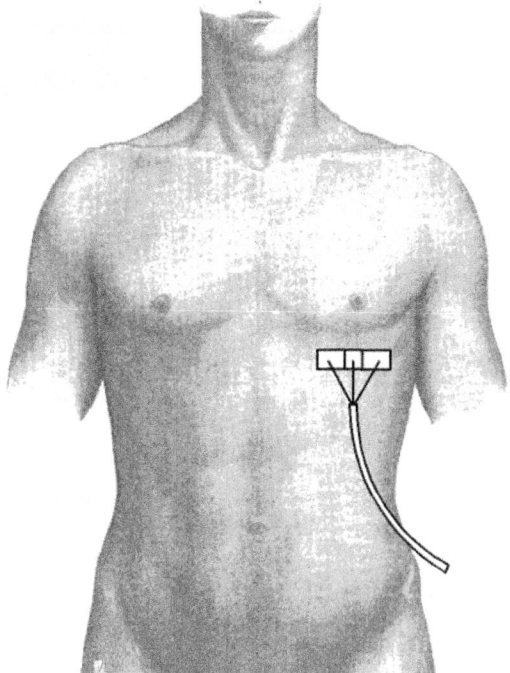

Figure 6.3 Placement of the probe

7 2.5 WATTS AMPLIFIER

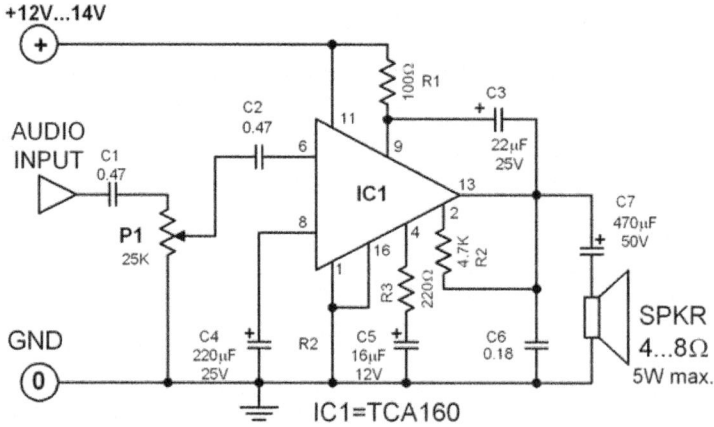

+12V...14V

AUDIO
INPUT

GND

C2 0.47

C1 0.47

P1 25K

IC1

R1 100Ω

C3 22μF 25V

C7 470μF 50V

R2 4.7K

R3 220Ω

C4 220μF 25V

R2

C5 16μF 12V

C6 0.18

SPKR
4...8Ω
5W max.

IC1=TCA160

Diagram 7.0 2.5 Watts Amplifier

There are many cases where you desperately need a simple to build and inexpensive amplifier that delivers moderate power. This compact amplifier delivers just enough power output so that you can hear the audio signal coming from any device. It delivers up to 2.5 watts audio output.

The heart of the amplifier is a single compact IC which makes the circuit simple to build and eventually troubleshoot. The IC used in this circuit is a TCA160 which integrates a full amplifier circuit including preamp and driver stages. The supply voltage range is flexible from 6V up to a maximum of 14V. The IC is normally heatsinked to avoid being damaged from overheating.

The actual power output of the IC depends on both the supply voltage and the speaker's impedance. It can be found using the formula shown.

Formula for P

$$P = \frac{V_b2}{8RI}$$

where Vb= actual supply voltage and RI= actual speaker impedance.

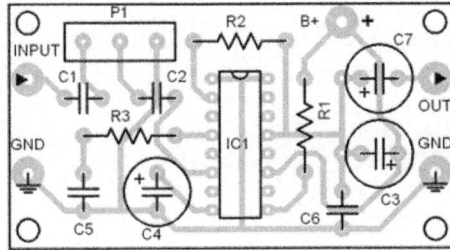

Figure 7.0 Parts Placement Layout

Figure 7.1 Printed Circuit Layout

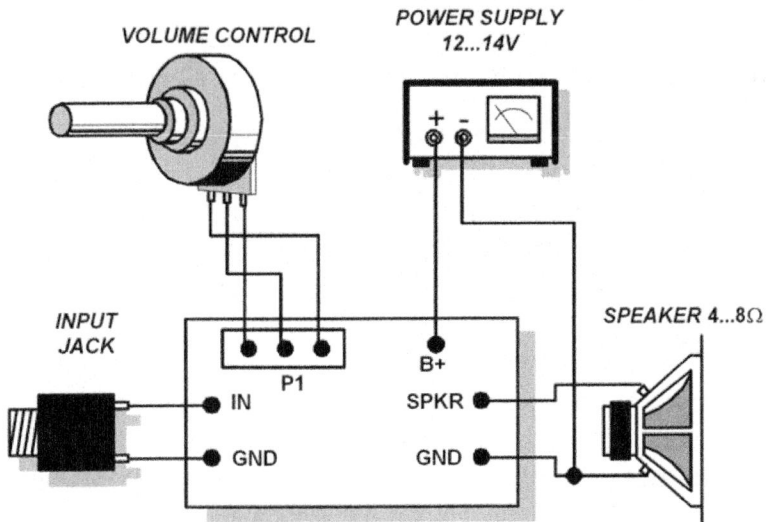

Figure 7.2 External Wiring Layout

8   AUDIO SQUELCH

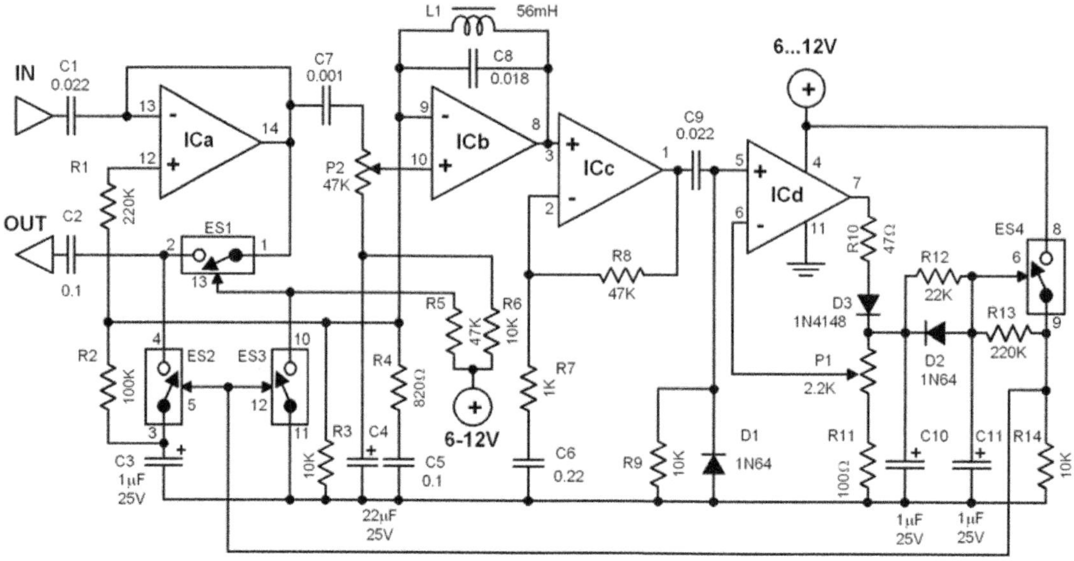

Diagram 8.0   Audio Squelch

ICa...d = LM324
ES1...ES4 = 4066

This squelch circuit is used in communication receivers to block out the audio (and noise) when no signal is being received. It works as a signal-to-noise ratio controlled squelch. It is originally designed for narrow band FM receivers. This circuit functions based on the fact that the receiver produces more noise when it detects no transmitted signal. When this noise exceeds a certain preset level, the circuit cuts off the connection between the demodulator and the audio amplifier input.

You can see in the block diagram how the circuit works. The output signal of the demodulator is no more connected to the input of the amplifier but instead enters the buffer ICa. The signal is then passed on to the bandpass filter ICb. The filtered signal is amplified by IC3 and rectified by ICd. The noise that is able to pass through ICb is amplified, rectified, and used as control signal for the electronic switch ES4. Switch ES4 in turn controls the switches ES1 and ES2.

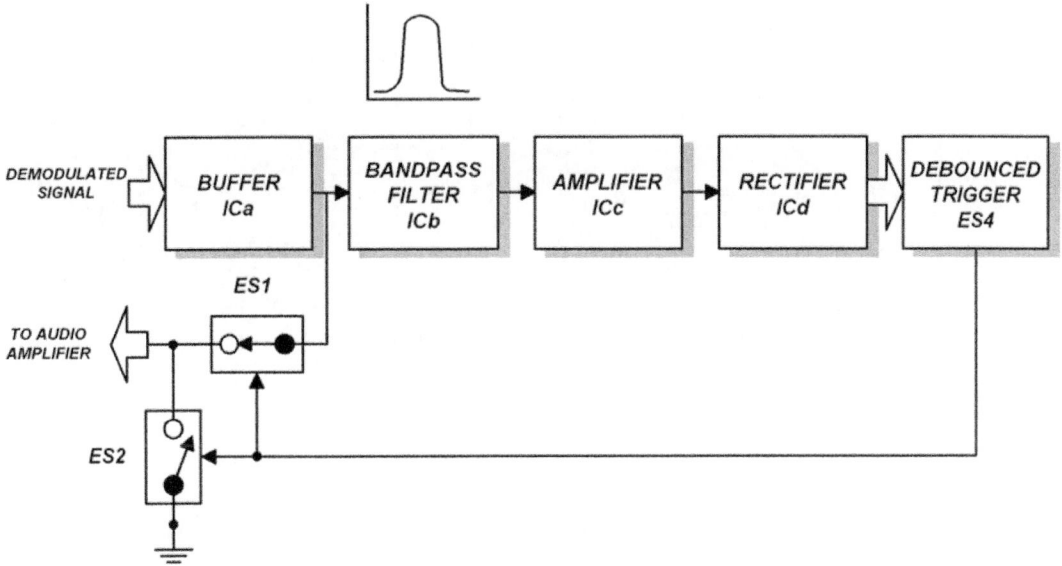

Diagram 8.1   Block Diagram of Audio Squelch

When the noise level is below the threshold, ES1 is closed and ES2 is open so that the output of the demodulator is fed to the audio amplifier. Otherwise, when the noise level exceeds the threshold, ES1 opens and ES2 closes. The connection between the demodulator's output and the audio amplifier is cut off.

In using the circuit, the connection between the demodulator and the audio amplifier must be broken, and diverted to the proper terminals in the squelch circuit as shown in the Diagram 8.2. Potentiometer P2 adjusts the amplitude of the signal entering ICa. P3 sets the gain of rectifier ICd.

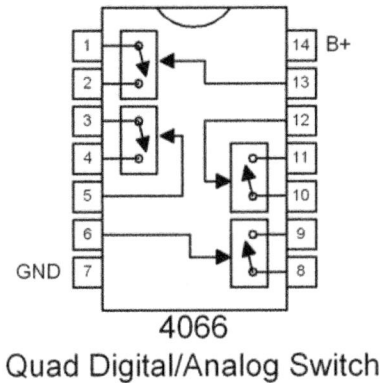

4066
Quad Digital/Analog Switch

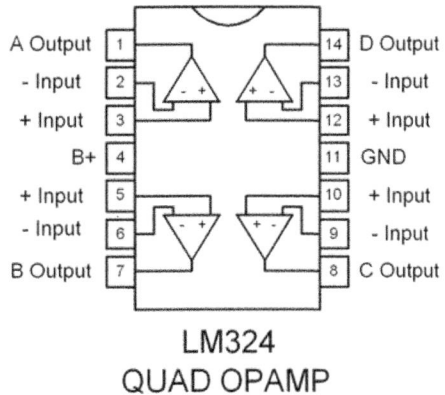

LM324
QUAD OPAMP

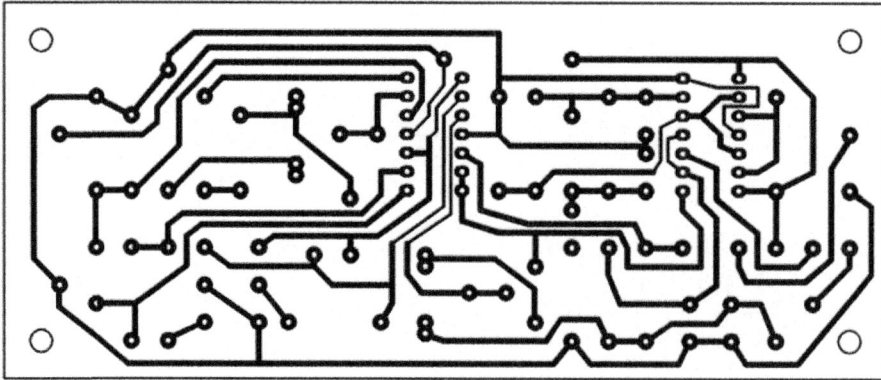

Figure 8.0   Printed Circuit Layout

Figure 8.1   Parts Placement Layout

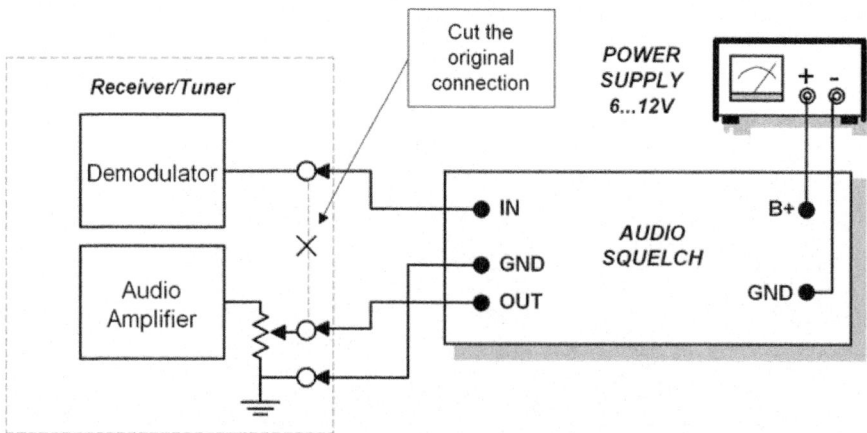

Diagram 8.2   External Wiring Layout for Audio Squelch

9 LOW NOISE PREAMP

A preamplifier circuit with a very low noise characteristic can be built by simply combining a FET transistor with a normal bipolar transistor.

The input impedance of the resulting circuit is almost the same as the gate impedance of a single FET transistor - around 1 megaohm. The output impedance at the other end is about 1 kiloohm.

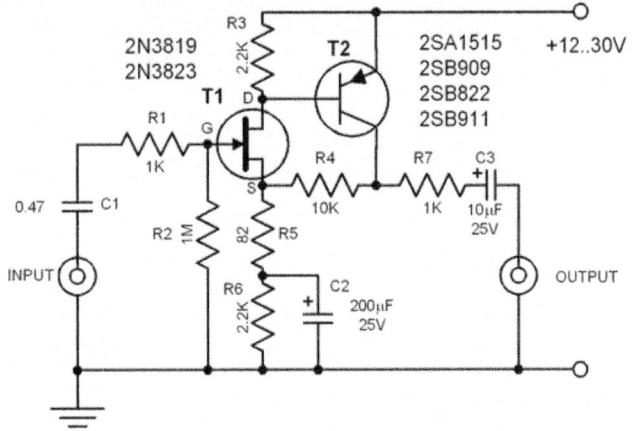

Diagram 9.0 Low Noise Preamp

The frequency of this preamplifier is linear (-3 dB) between 10 Hz and 450 kHz and (-1 dB) between 20 Hz and 200 kHz. The amplifying factor is around 100.

This circuit can be powered from 12 to 30 volts without significant deterioration in its amplifying characteristics.

2SB822
2SB909
2SB911

B C E

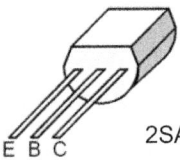

2SA1515

E B C

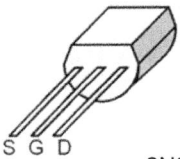

S G D

2N3819

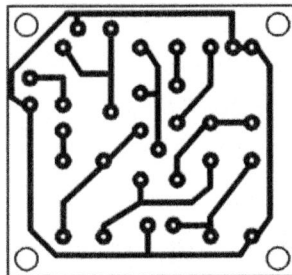

Figure 9.0
Printed Circuit Layout

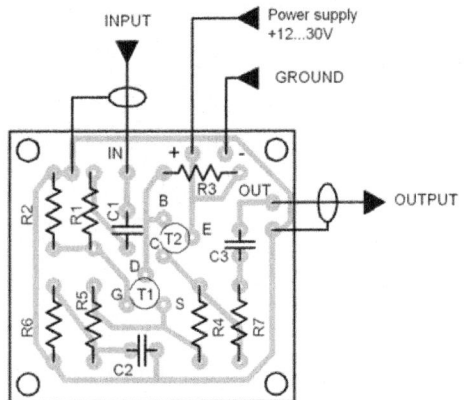

Figure 9.1
Parts Placement Layout
& External Wiring

10   MINI AUDIO AMPLIFIER

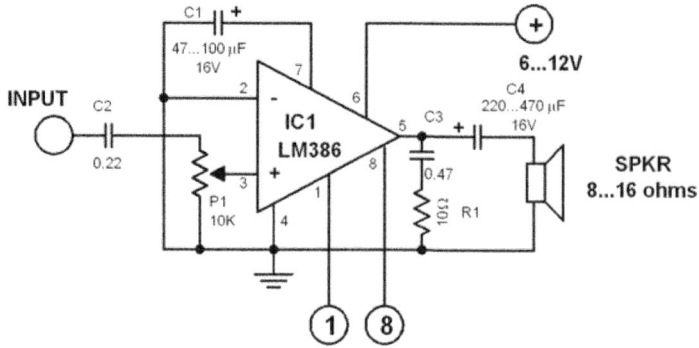

Diagram 10.0 LM386 Audio Amplifier

The integrated chip LM386 is a low power audio frequency amplifier requiring a low level power supply (most often batteries). It comes in an 8-pin mini-DIP package. The IC is designed to deliver a voltage amplification of 20 without external add-on parts. But this voltage gain can be raised up to 200 (V_u = 200) by adding external parts.

The external parts shown on Diagram 10.1 can be selected to get the desired gain. Circuit A will give a voltage amplification of 200. Circuit B will give a gain of around 50. The circuit C is not for voltage amplification but will raise the bass level by about 5 dB. Take note also that the circuit C is to be connected between pins 1 and 5 of the IC.

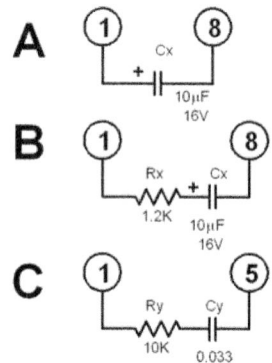

Diagram 10.1 External parts

The power output is around 550 mW at 16 ohm speaker impedance. This audio frequency amplifier is ideal for small battery powered devices.

If you use the external circuit A, replace the Rx with a jumper wire in the pcb. If you use the external circuit C, solder the additional resistor and capacitor to the pcb points labeled Ry and Cy, see Figure 10.1.

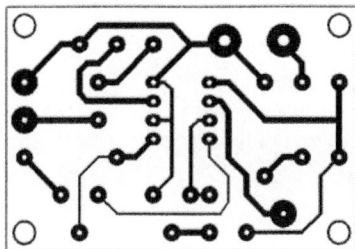

Figure 10.0
Printed Circuit Layout
for the Mini Audio Amplifier

Figure 10.1
Parts Placement Layout
for the Mini Audio Amplifier

11   6.5 WATTS AMPLIFIER

Diagram 11.0   6.5 Watts Amplifier

A medium powered amplifier for universal audio amplification purposes can be constructed with the monolithic amplifier TCA940E. This IC has negligible harmonics, low distortion, and a built-in short circuit protection. The IC's own heatsink must be soldered to the PCB's ground. Although its own heatsink is usually enough for most applications, it is a good idea to physically and thermally connect it to a large copper plate of the PCB. An area of around 4 cm^2 to 6 cm^2 is sufficient.

Technical Data

Power supply = 12 - 30 V
Power output = 6.5 W with 20V supply (8 ohm load)
           5.4 W with 18V supply (8 ohm load)
Frequency response= linear from 40 Hz - 20 kHz
Distortion factor (at 50mW to 3.5W) = 0.2%
Input sensitivity   =   110 mV
Input Impedance =   100K

Figure 11.0   Printed Circuit Layout

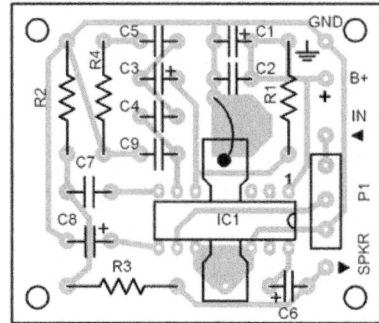

Figure 11.1   Parts Placement Layout

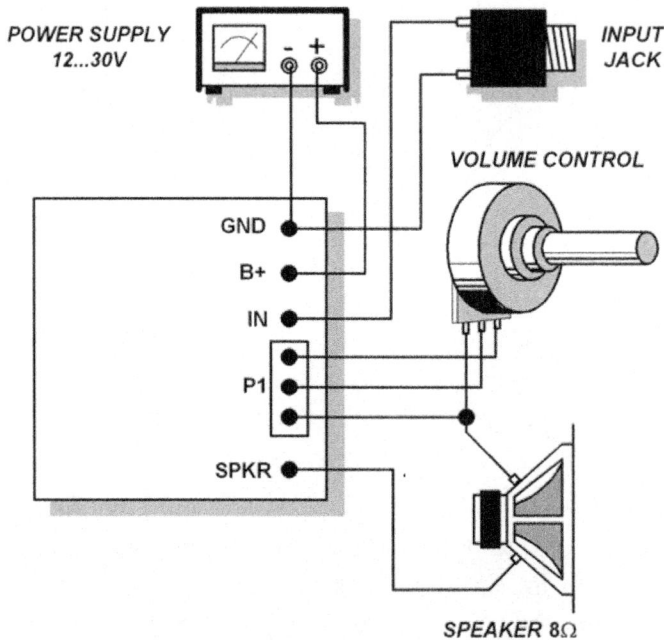

Figure 11.2   External Wiring Layout

12 DYNAMIC MIC PREAMP

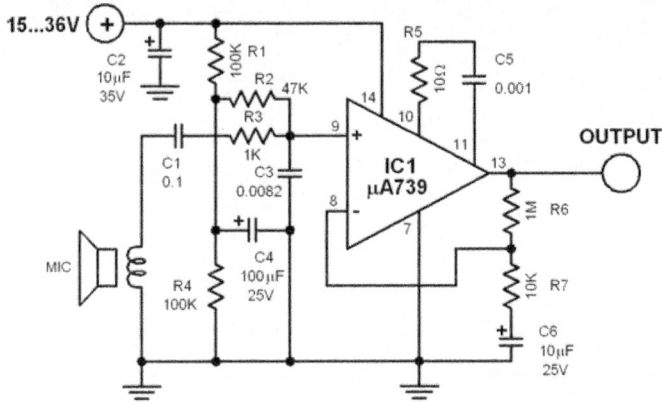

Diagram 12.0 Dynamic Mic Preamp

The above mic preamp uses the low noise µA739 IC. The circuit is an example of how a good preamp can be designed for dynamic microphones. The IC houses two identical integrated preamp circuits. The second preamp is used in identical manner for the second channel of the stereo microphone. Diagram 12.1 shows the pin numbers (in brackets) for the second identical channel. All external parts are identical to those shown in Diagram 12.0 The non-inverting input is biased at about 50% of the power supply. This bias voltage is set by the voltage divider circuit R1 and R4. The point between R1 and R4 is used commonly for both channels.

The unwanted HF signals coming from the microphone are filtered out by the RC-circuit composed of R3/C4. Frequency compensation is done by the R7/C6 circuit. The values of R7 and C6 were designed to avoid oscillation at the amplification level of 100. The input impedance is about 47K. This means that a normal dynamic microphone gets connected to a high impedance preamp which in turn produces good results. The output impedance is about several hundred ohms.

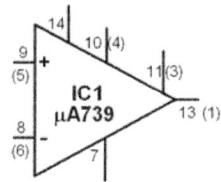

Diagram 12.1 Pin designations for the other preamp inside the IC

The maximum peak-peak output voltage is about several volts lower than the supplied power. The frequency range is from 20Hz to 20kHz (-3dB). The upper cutoff frequency is about 80 kHz when the low-pass filter is removed from the circuit. The IC shown can be replaced with TBA231 or SN76131 without changing the external circuit.

13   FET AUDIO MIXER

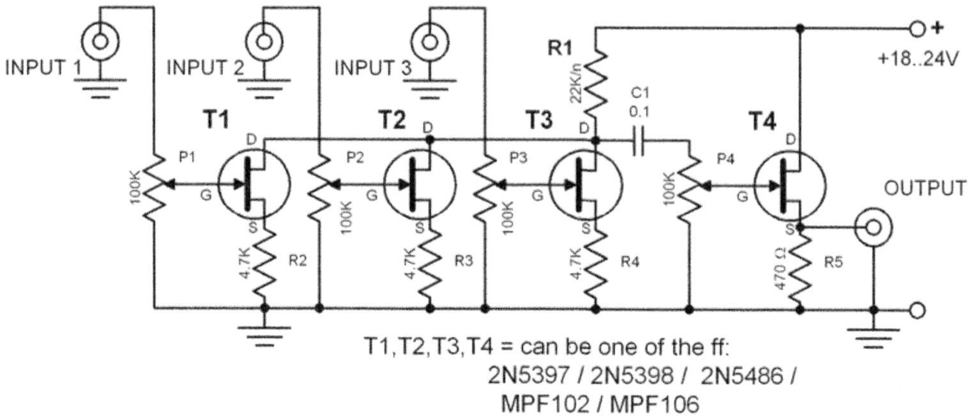

T1,T2,T3,T4 = can be one of the ff:
2N5397 / 2N5398 / 2N5486 /
MPF102 / MPF106

Figure 13.0   FET Audio Mixer

Although FETs are originally designed for high frequency applications they can also be used for audio frequencies. In fact, they perform excellently in this area. The audio mixer featured here is an example of the FET's versatility. This mixer needs only four FETs and performs very well.

The mixer's input impedance is determined by the resistance of the input potentiometers used in the circuit since the impedance of the FETs itself is very high.

The number of inputs that can be connected to the circuit is practically unlimited as long as the value of R1 is chosen according to the above given formula.

Formula for R1
$$R1 = \frac{22K}{n}$$
where **n** = the number of inputs

The frequency response of the mixer is from 20 Hz to 80 kHz and linear within 3 dB.

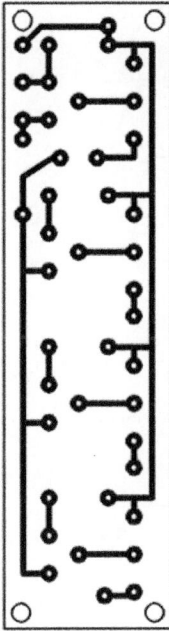

Figure 13.0 Printed Circuit Layout

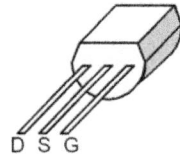

Figure 13.1 Parts Placement Layout

Parts List:

Resistors:
R1   = 22K/n (see text)
R2,R3,R4   = 4.7K
R5   = 470 Ω
P1,P2,P3,P4 = 100K Potentiometer
All resistors are ¼ watts unless
otherwise specified.

Capacitor:
C1 = 0.1µF/50V ceramic

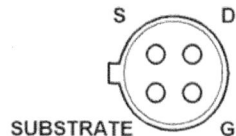

2N5486
MPF102

D S G

2N5397
2N5398
MPF106   SUBSTRATE

Transistor equivalents
NOTE: Transistors in bottom view

14 MIC PROCESSOR

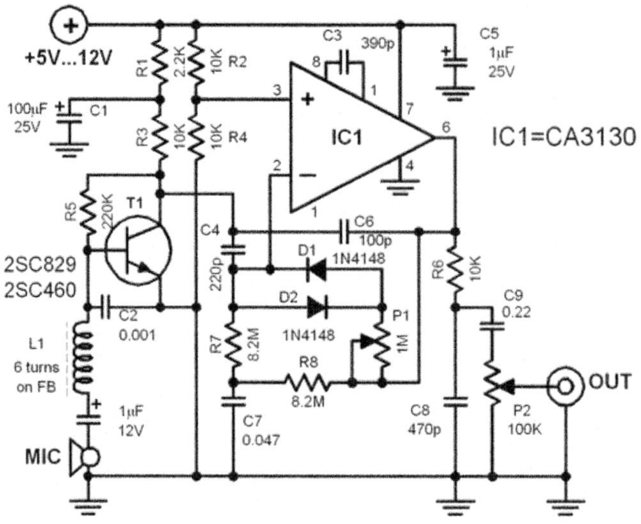

Diagram 14.0 Mic Processor

Audio processors are usually used in paging systems, in wireless intercom and the likes to amplify the microphone signal to a certain level. This can be done by using either a compressor or a limiter circuit. Although a compressor has lower distortion characteristics, it is a very complicated circuit. The limiter is simpler to construct, but it has a relatively high distortion level. Intermodulation distortion is high in a limiter circuit and in order to effectively use a limiter, you have to suppress the intermodulation interference as much as possible. This can be done by automatically changing the limit frequency according to the strength of signal input. The circuit featured here does just that.

This electronic circuit has an amplifier with a very high input impedance. When the input signal is still low, the diodes do not conduct yet. In this situation the limit frequency is still dependent on R1 and C1. Once the diodes conduct, (it happens when the input signal has increased), the input impedance of the amplifier decreases thereby shifting the limit frequency to a higher value. The lower frequencies are then amplified less thus making the audio signal more understandable. The signal processed this way is much better than the one which is just simply "clipped". This circuit is also applicable to process music signals.

The values of C6, C7 and C8 are given in the following table according to the application of the circuit.

	C6	C7	C8
music applications	-	47n	470p
voice applications	100-220p	0 - 4n7	4n7

15 SPEAKER PEAK INDICATOR

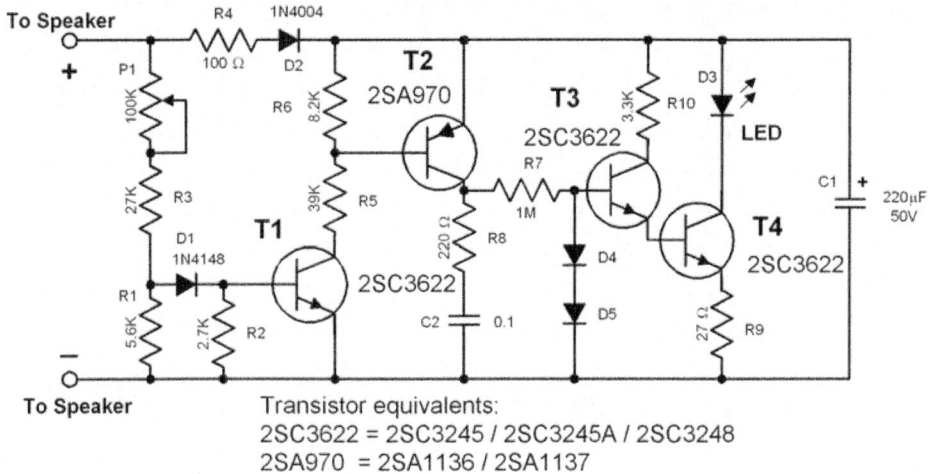

Diagram 15.0 Speaker Peak Indicator

Transistor equivalents:
2SC3622 = 2SC3245 / 2SC3245A / 2SC3248
2SA970 = 2SA1136 / 2SA1137

Modern speaker boxes are relatively insensitive to overdriving signals, however it is still important to limit the power driving it to avoid the clipping of the audio signal. A broken sound from an overdriven loudspeaker is not only annoying to the ears but also the loudspeaker system itself might be damaged by the continuous uncontrolled signal peaks.

This peak indicator circuit is a very useful aid in detecting the driving limits of a speaker system. It can be connected directly to the existing speaker wirings and it needs no extra power. It can detect very short voltage overswings and therefore provides reliable means of determining the driving limits of a speaker.

Simply listening to the speaker is inaccurate because short time distortions are difficult to detect.

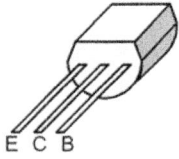

2SA970
2SA1136
2SA1137
2SC3622
2SC3245
2SC3248

*Transistor
equivalents*

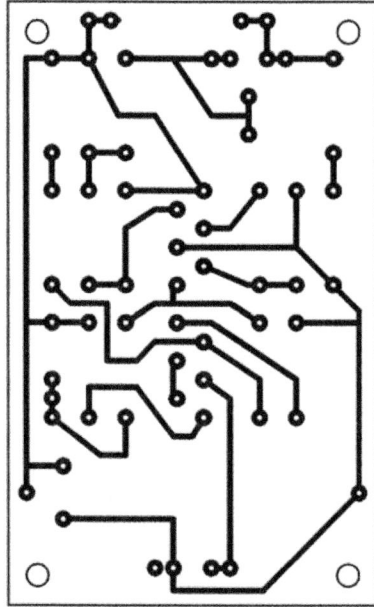

Figure 15.0
Printed Circuit Layout

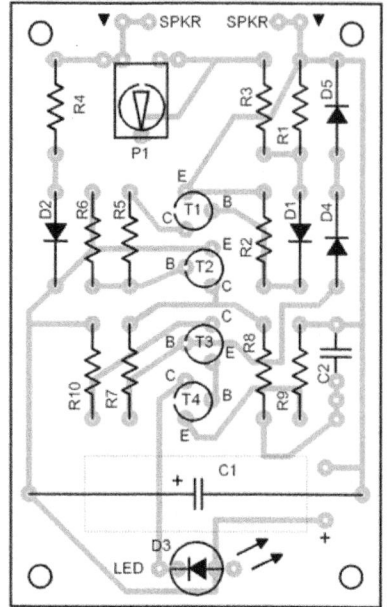

Figure 15.1
Parts Placement Layout

   The threshold level of the peak indicator can be set for speakers ranging from 15 watts to 125 watts with 8 ohms impedance. For 4 ohms speakers, the indicator can be set from 30 watts up to 250 watts. By testing speakers, occasional blinking of the LED does not mean danger but when the LED blinks very often then it is advisable to reduce the volume of the amplifier.

   The circuit operates this way: During operation the signal charges C2 through R1 and D1. In standby periods, all transistors are turned off and no current flows through the LED. A sample of the signal flows through P1 and enters T1. P1 controls the threshold level of T1. Once the signal sample exceeds the threshold level, T1 and T2 switch on and charges C1 as a result; T1 conducts and switches T4 on. When the signal goes down, T1 and T2 turn off but since C1 discharges slowly, T3 and T4 remain conducting for about one second longer causing the LED to also light up longer. This technique has the advantage of indicating short time signal overswings which are normally not detectable.

   In constructing the circuit, it is advisable to use high luminance LEDs with a diameter of not less than 3mm.

16 AUTOMATIC VOLUME CONTROL

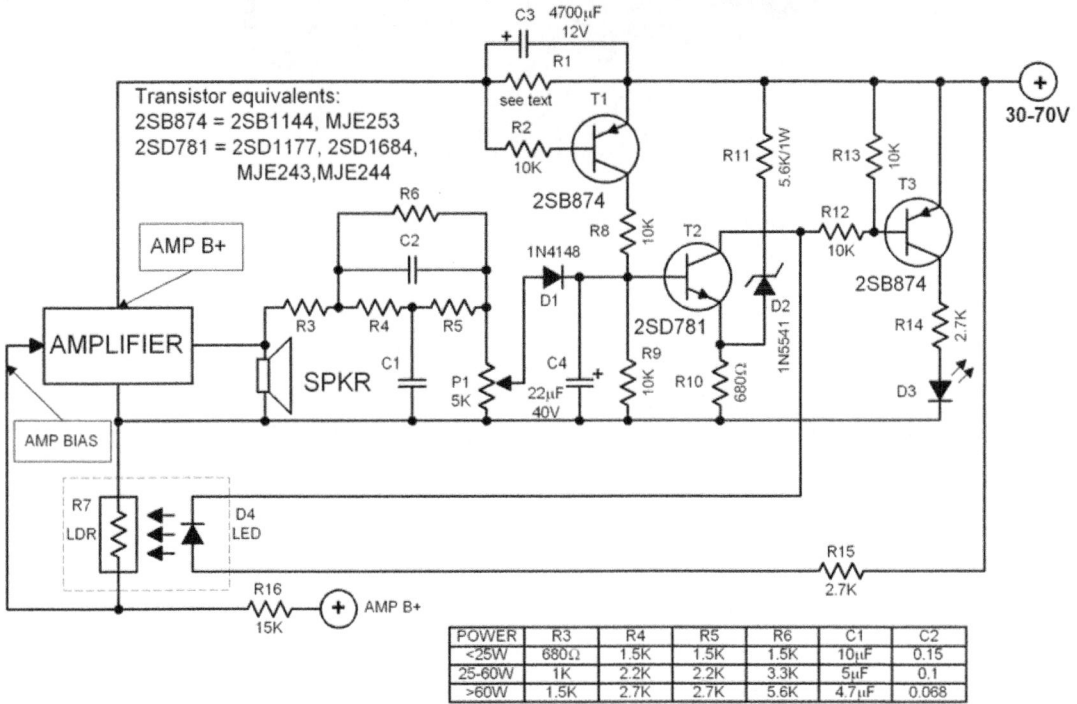

Transistor equivalents:
2SB874 = 2SB1144, MJE253
2SD781 = 2SD1177, 2SD1684,
MJE243, MJE244

POWER	R3	R4	R5	R6	C1	C2
<25W	680Ω	1.5K	1.5K	1.5K	10μF	0.15
25-60W	1K	2.2K	2.2K	3.3K	5μF	0.1
>60W	1.5K	2.7K	2.7K	5.6K	4.7μF	0.068

Table 16.0   Filter Network Values

Diagram 16.0   Automatic Volume Control

This circuit limits the power output of an amplifier (10 to 100 watts) according to the combination of its output level, frequency, current consumption and supply voltage. The values are of course dependent on the actual amplifier used. This circuit can be adapted to the individual amplifier type very easily. It also accepts a very wide range of power supply voltages from 30 volts to 70 volts!

For example the current consumption of the amplifier is sampled through R1. As you know very well, the silicon transistor T1 starts to conduct when its base voltage reaches around 0.56 volts. Once the current consumption of the amplifier produces a voltage drop of 0.56V at resistor R1, the circuit (actually transistor R1) starts to limit the amplifier's gain. You have to compute the value of R1 basing on the current consumption of your amplifier. Just use the Ohm's law.

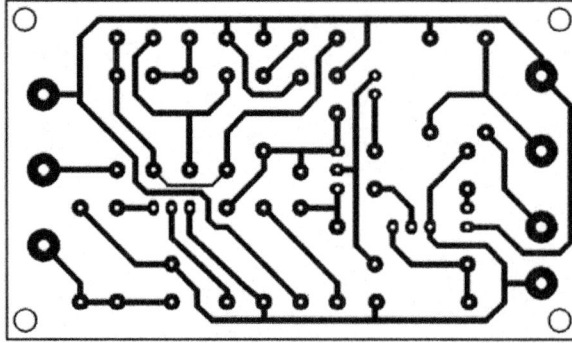

Figure 16.0   Printed Circuit Layout

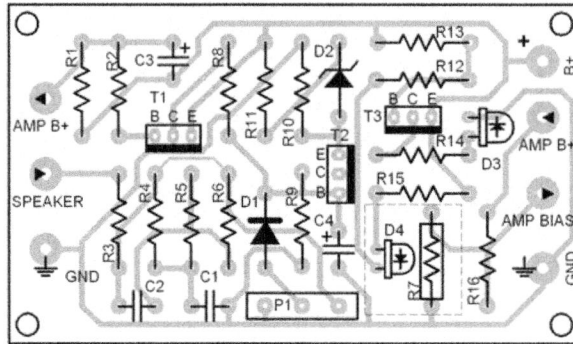

Figure 16.1   Parts Placement Layout

The actual gain control of the amplifier is done by the LDR/LED combination shown in the circuit. The LDR/LED combination must be housed inside a lightproof box. The AVC is also connected either to the loudspeaker or to the output terminals of the final amplifier. A filter network is connected right after the speaker.

E C B

2SB874	2SB1144
2SD781	2SD1177
2SD1685	MJE243
MJE244	MJE253

The sample signal passes first through the filter before it is processed by the main circuit. The component values are dependent on the frequency and output power. The component values for the filter are listed in Table 16.0. The limit threshold can be set through the trimmer P1. If you want to detect the signal peaks only, remove the filter circuit. In such case the circuit is frequency independent.

17 AUDIO MIXER

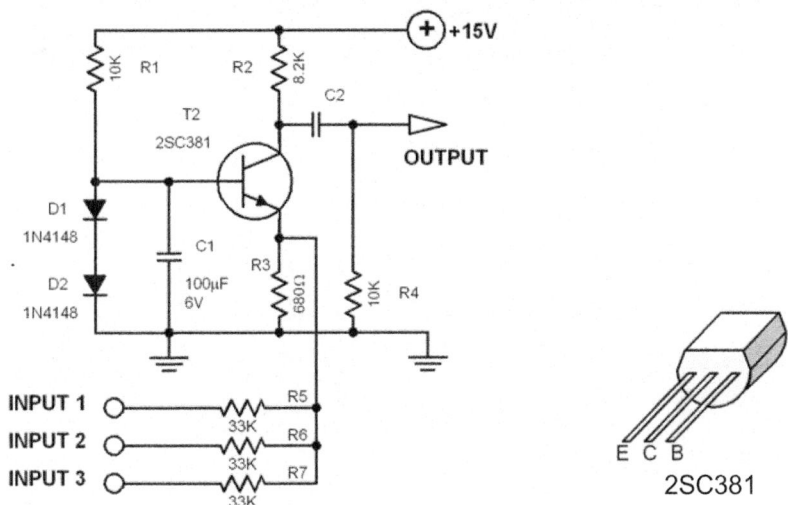

Diagram 17.0 Audio Mixer Circuit

In an amplifier circuit with base driven transistor and with its emitter being current controlled, most of the driving current flows through the collector away. Using the values in the circuit shown in diagram 17.0, the collector current will be about 1 mA. At 15 volts power supply, the input resistors should be 33K. Additional input lines can be connected to the emitter line. Each added input must be series limited by the 33K resistor.

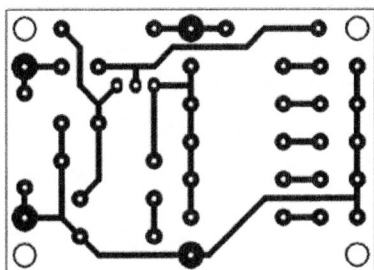

Figure 17.0
Printed Circuit Layout

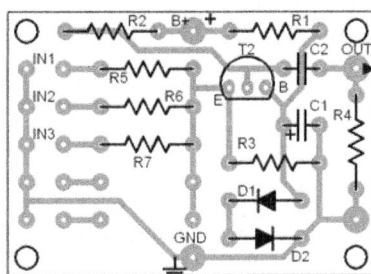

Figure 17.1
Parts Placement Layout

18 LOW NOISE MIC AMP

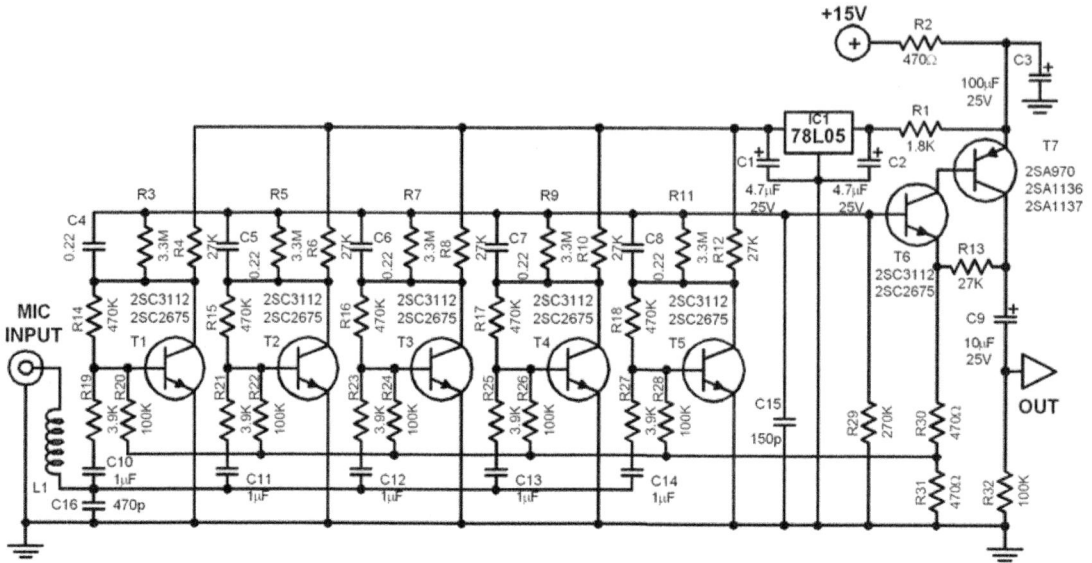

Diagram 18.0 Low Noise Mic Amp

This microphone circuit is originally designed for recording sounds from very far distances using microphones with parabolic reflectors. Its noise level is around 7 dB and has a distortion factor of around 1%. This value is less than the distortion factor of a tape recorder. A 130μV input signal gives an output level of 60mV. This figure shows that the microphone amplifier is very sensitive. The amplification factor is around 475. The circuit can also accept input levels up to 8mV. Its bandwidth is from 20 Hz to 45 kHz.

Coil L1 is several turns of magnet wire in a ferrite core. This microphone amplifier is about 12 dB better than the mic amp of a good tape recorder. Of course, this circuit can also be used in standard music recording. The quality of the recorded music is much better than the one recorded with an ordinary amplifier.

To obtain the best results, the resistors used in this circuit are of metal film type.

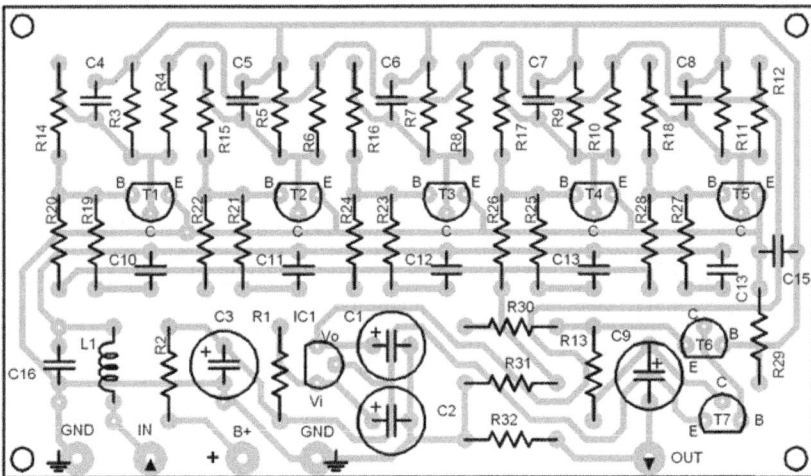

Figure 18.0   Printed Circuit Layout

Figure 18.1   Parts Placement Layout

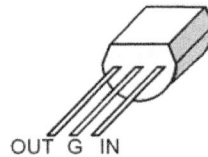

E C B

2SA970   2SC3112
2SA1136   2SC2675
2SA1137

OUT G IN

78L05

19 CASSETTE PREAMP

Diagram 19.0 Cassette Preamp

This circuit amplifies the very weak signals picked up by a cassette magnetic head. This preamp is highly insensitive to noise. The usual click or popping sound at the start of every playback is absent since the preamp inputs are directly connected to the head coils without a coupling capacitor.

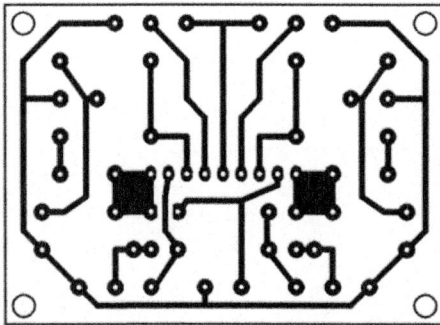

Figure 19.0 Printed Circuit Layout

Figure 19.1 Parts Placement Layout

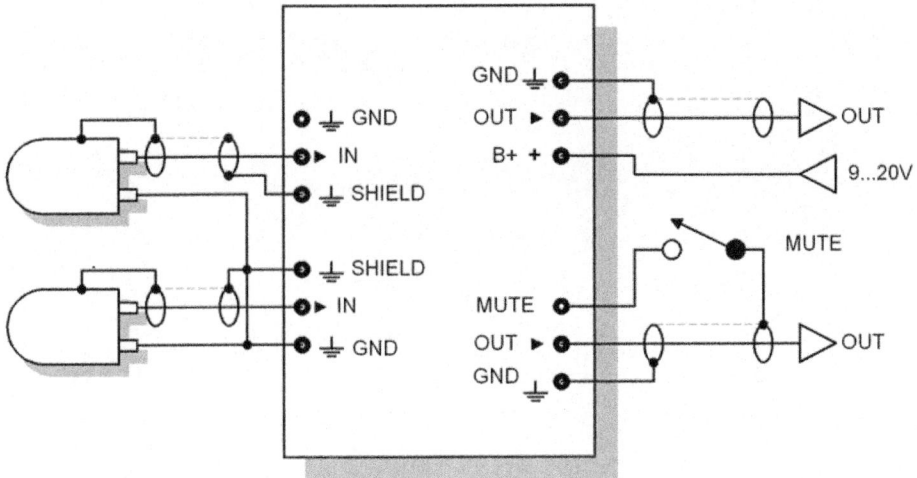

Figure 19.2 External Wiring for 2 Mono Heads

Figure 19.3 External Wiring for single Stereo Head

Technical Data	
Supply Voltage=	7.5V - 23 V(typical 8.5V)
Standby amplification=	90 dB
Current consumption=	7.5 mA
Output power =	1.5 Vrms maximum
Distortion factor=	0.05%
Input resistance=	over 200K
Output resistance=	less 1K

20 ELECTRONIC SWITCHBOARD

U1...U4 = 4066 = IC3...IC8

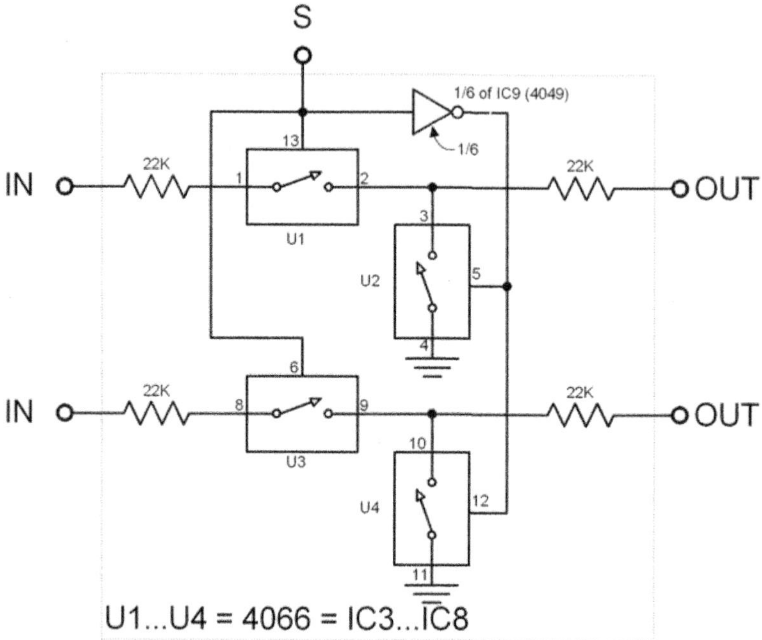

Diagram 20.0 Electronic Switchboard

You know the problem only too well: signal sources of all kinds, cable salad, connectors. Well, if electronics brought this problem why not use electronics to solve it as well. Right! ... and the solution is an electronic signal switchboard. This switchboard circuit is modifiable according to needs. With this circuit you can say goodbye to cable salad.

The circuit featured here can be easily adapted to your needs. It has one free definable input, two tape recorder inputs and one echo effect input. All four inputs can be connected to any of the three outputs (two tape recorder outputs and one echo effect output). The changing of the connections is done by simply flipping any of the six toggle switches that controls the electronic analog switches.

The heart of the circuit is composed of six electronic analog switch modules (ES1 up to ES6) that you see on the left part of Diagram 20.1. Detailed circuit design is shown in diagram 20.0. One ES module uses one 4066 analog switch and 1/6 of the 4049 inverter of IC9.

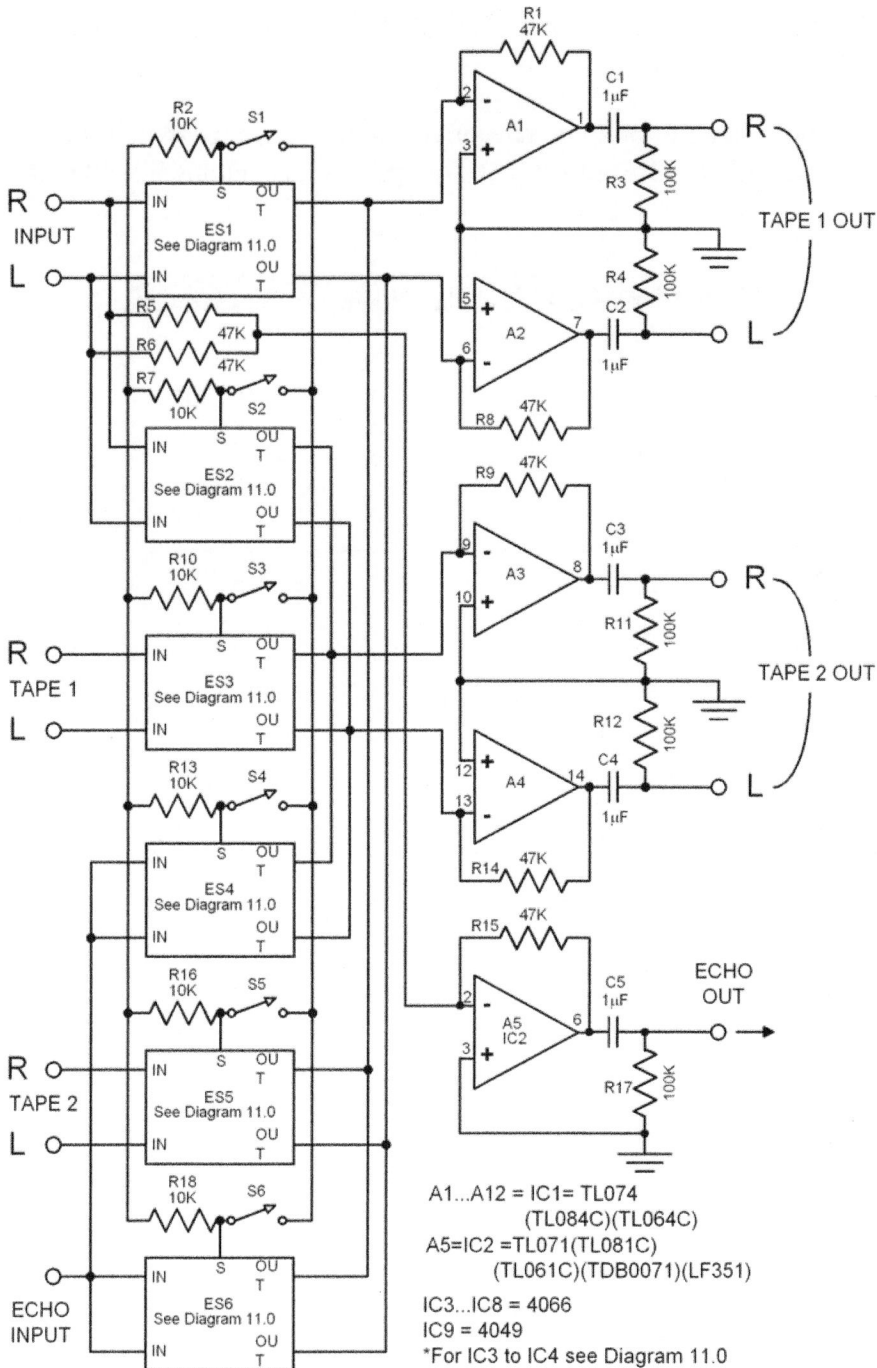

Diagram 20.1 Electronic Switchboard

The inner function of each of these switches is shown in Diagram 20.0. The inner switches work much like ordinary relays. They are controlled by a voltage that is applied to the S-input. Each signal channel needs two electronic switches. When the voltage applied to S-input is positive, all inner switches will open, preventing the input signal to get through. When the S-input sees a negative voltage, the inner switches will close. The output signal is finally amplified by IC5. The amplification factor is variable through R11.

Technical Specifications

Channel separation:	75 dB
Signal/noise ratio:	more than 100 dB
Distortion factor:	less than 0.01 %
Current consumption:	approx. 20 mA
Supply voltage:	two 9 Volt blocks

Diagram 20.2 Supply connection to IC's

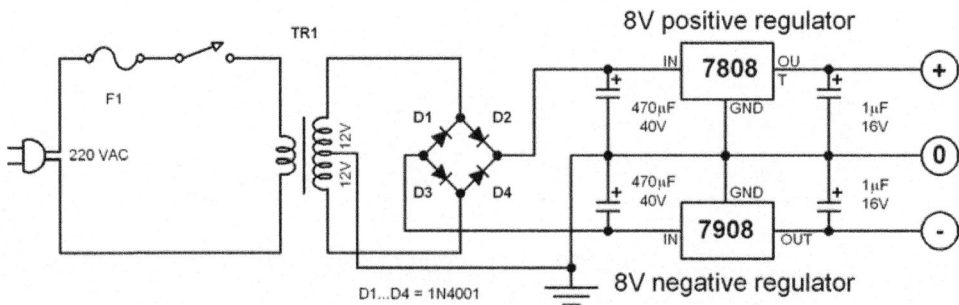

Diagram 20.3 Power Supply (Electronic Switchboard)

21 TUNABLE FILTER CIRCUIT

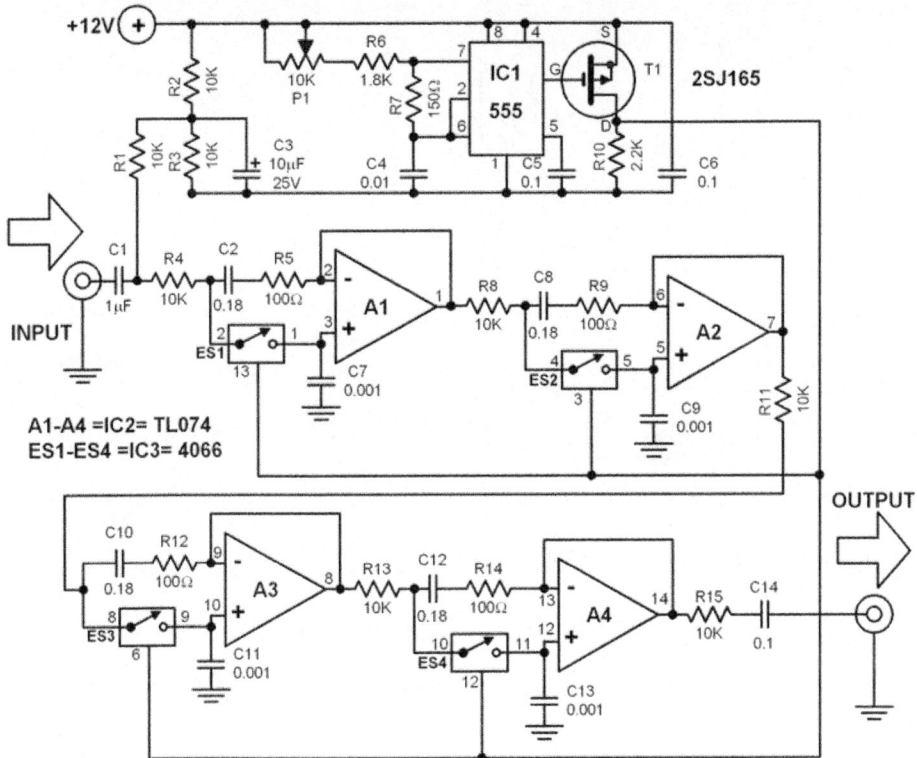

Diagram 21.0 Tunable Filter Circuit

This circuit uses a clock generator to control four analog CMOS switches. The switches in turn vary the input resistance of each op-amp. These analog switches are controlled by a 555 clock generator with a duty cycle that is variable from 1:10 up to 100:1. If the analog switch is closed, the input resistance is around 60 ohms and when it is open, the resistance is almost infinite. If, for example, the switch is clocked with a duty cycle of 0.5, the resulting input resistance is $1/(0.5/60) = 120$ ohms. If it is 0.25, the resistance is 240 ohms. It shows that every half of the duty cycle results to a doubling of the input resistance.

The frequency of the clock signal must be several times higher than the highest audio frequency to be processed to avoid hearing the resulting interference between the clock and the audio signals. The amplification is around 40 and is dependent on the clock frequency.

22 SUBWOOFER FILTER

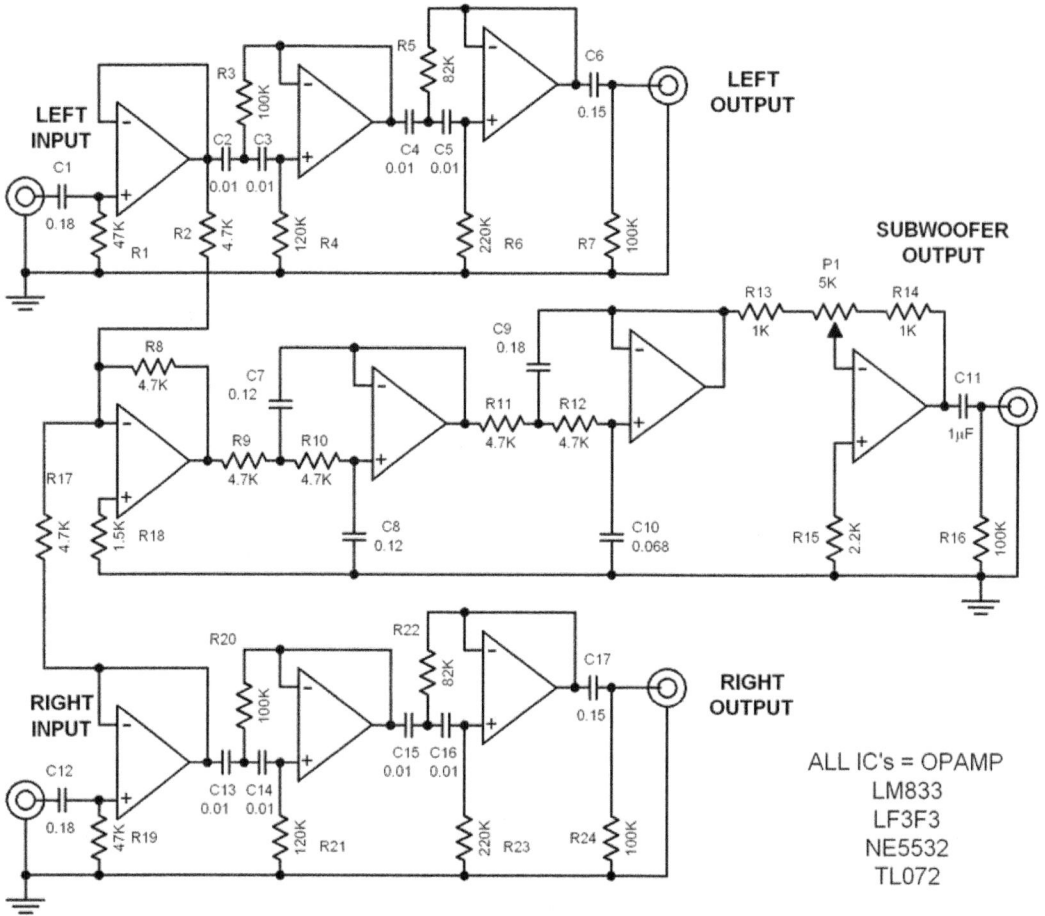

Diagram 22.0 Subwoofer Filter

If you are interested in experimenting with audio circuits in the subwoofer range, this circuit is for you. In the subwoofer range, all audio frequencies below 200 Hz can be fed to a single speaker box (mono) since the human directional perception of sound diminishes at this frequency range. In short, you don't need to use (or buy) two bass speakers for both of the stereo channels. You use only one bass speaker. The normal stereo signals above 200 Hz can be fed to two satellite speaker boxes.

The electronic circuit shown in diagram 22.0 is basically an active filter. It is a 24 dB octave filter with a Bessel characteristic and cutoff frequency of 200 Hz.

How does the circuit work: Op-amps A1 and A2 buffer the signals coming from both right and left channels. Op-amp combinations A3/A4 and A9/A10 function as the highpass filters for both stereo channels. The outputs are then connected to the final amplifiers of the satellite boxes. Signals from both left and right channels are fed to the op-amp A5. Op-amps A6/A7 function as the lowpass filter. Op-amp A8 works as the output amplifier for the subwoofer signal. The signal level can be balanced between the subwoofer and the satellite lines.

The power needed for this circuit must have a symmetrical output. The op-amps can have either JFET or bipolar inputs.

23 PREAMP w/ EQUALIZER

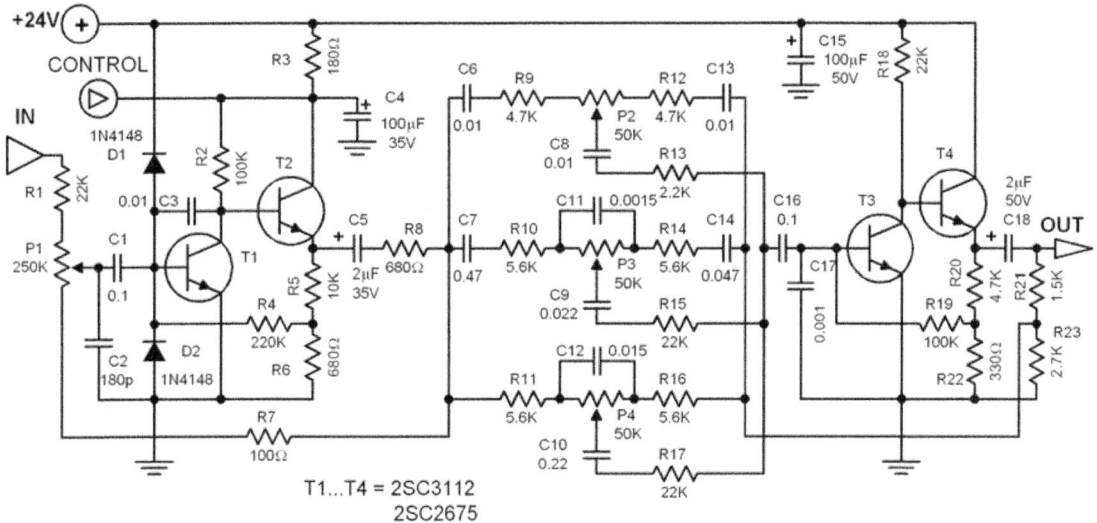

Diagram 23.0 Preamp with Equalizer

This universal preamp can be used for different applications. It enables the amplification of many types of signals. The input is protected from overdrive by the diode combination D1 and D2. The output impedance is low which enables it to be connected to any type of final amplifier without problems. The preamp is composed of two dc-coupled amplifier stages. A simple three-way equalizer is connected between these two amplifier stages. The amplification level is variable through P1.

24 40-WATTS AMPLIFIER

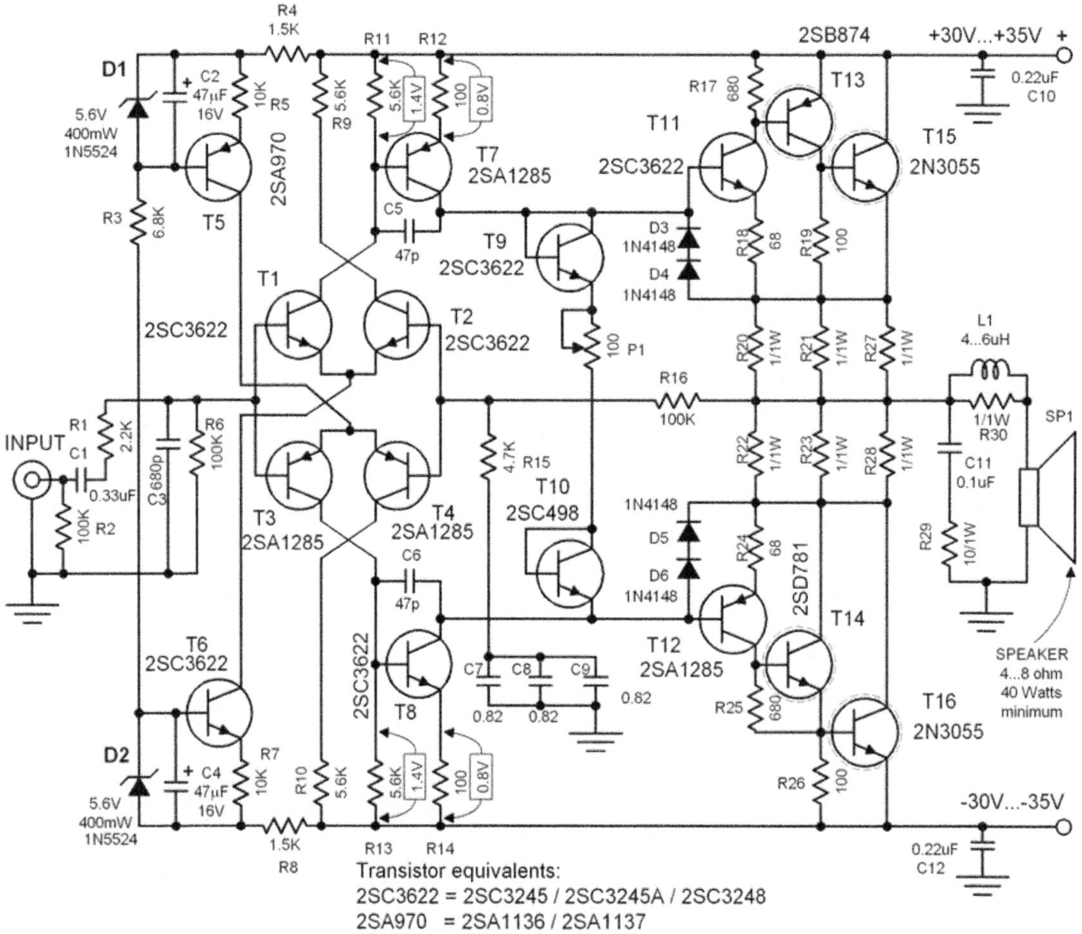

Transistor equivalents:
2SC3622 = 2SC3245 / 2SC3245A / 2SC3248
2SA970 = 2SA1136 / 2SA1137
2SA1285 = 2SA1285A
2SB874 = 2SB874-B / 2SB874-C / 2SB1144 / MJE253
2SD781 = 2SD2277 / 2SD1684 / MJE243 / MJE244

Diagram 24.0 40-Watts Amplifier

This amplifier is designed for real music lovers who are not satisfied with just listening to the music but also want to "feel" the music. The 40 watts power coming from this circuit delivers a „punch". The very common 2N3055 transistors are used as final amplifiers. This makes the acquisition of the needed components very easy.

At first glance, you might think that the design of this electronic circuit is not symmetrical because the final transistors are both NPN. But wait, if you look closely at the transistor combination T12-T14-T16, you will realize that it functions as a PNP transistor. The transistor combination T11-T13-T15 on the other hand functions as a NPN transistor. Since these two transistor combinations work together as a complementary pair, the circuit is symmetrical.

The rest of the circuit is designed in the same way. The transistor combination T1-T2-T6 provides a current source and functions as the upper half of a differential amplifier. The transistor combination T3-T4-T5 functions as the lower half. The transistors T7 and T8 work as drivers. The voltage drop at resistors R25...R30 is about 33 mV by 50 mA standby current.

Transistors T15 and T16 must be heatsinked. Be sure that the transistors are electrically isolated from the heatsinks. Transistors T13 and T14 must also be heatsinked. Transistor pair T9 and T11 must be thermally coupled, meaning they must have a common heatsink. Transistors T10 and T12 must also be thermally coupled.

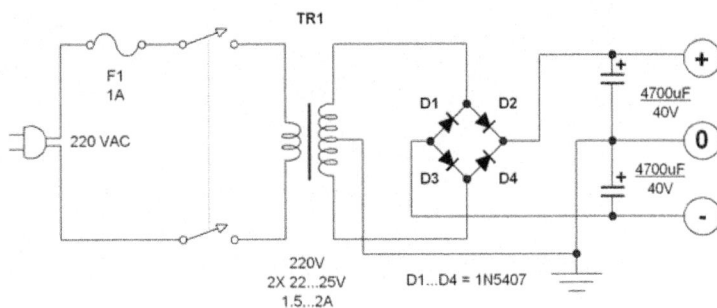

Diagram 24.1 Power Supply (40-Watts Amplifier)

Coil L1 must be self constructed: Wind 10 turns of magnet wire around R7. The ends of the magnet wire must be soldered to the terminals of R7.

TAKE NOTE

Only one channel of the circuit is shown here. To build a stereo system, you must build two identical circuits.

Technical Specifications	
Distortion factor:	0.01 % in 20 Hz - 20 kHz
Power output:	40 Watts (8 Ohms);
60 Watts (4 ohms)	
Input sensitivity:	800 mV for 40 W
	850 mV for 45 W
	700 mV for 60 W
	725 mV for 65 W
Frequency Response:	15 Hz - 100 kHz (+/- 1 dB)
Current amplification:	approx. 200,000
Standby current:	25 - 50 mA
Current consumption:	1 A by 40 W
	1.06 A by 45 W
	1.75 A by 60 W
	1.81 A by 65 W
** W = watts	

25 ELECTRONIC ORGAN

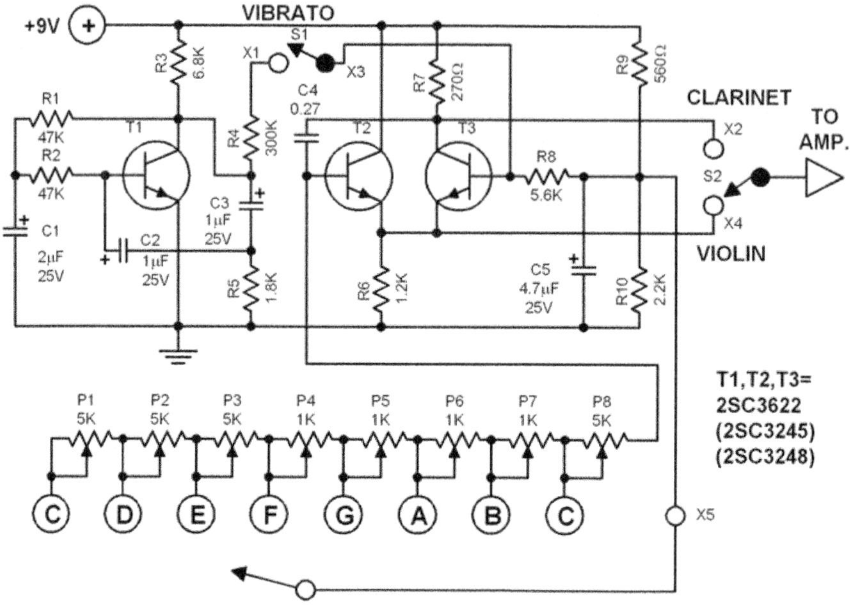

Diagram 25.0 Electronic Organ

This electronic organ is very simple to construct and can provide hours of entertainment particularly for children. The circuit is basically an emitter-coupled oscillator composed of T2 and T3. A square wave voltage can be sampled from the collector of T3(X2). This signal gives a clarinet character to the tone . Without the square wave signal, the sound produced by the emitters of T2 and T3(X4) has a violin character.

An additional vibrato signal can be added to this basic sound through switch S1. The frequency of the vibrato is approximately 6 Hz. Its amplitude is determined by the resistor R4. The value of R4 can vary from 100 up to 300K. Try experimenting with different values.

The keys can be made out of either metal plates or etched printed circuit. The trimmers P1 up to P8 adjust the pitch of each tone. The tones can be drastically changed by changing the value of C4.

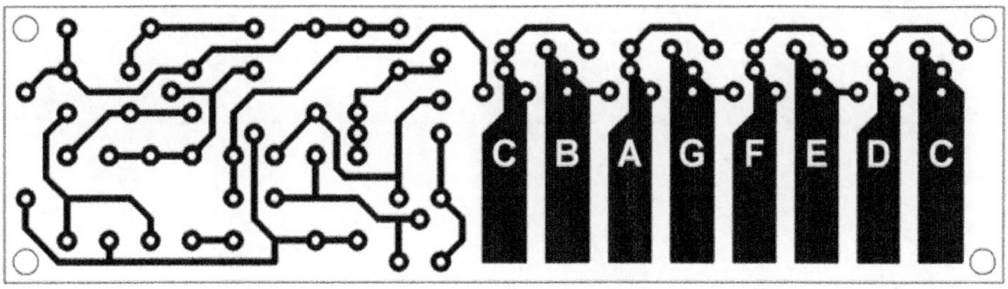

Figure 25.0 Printed Circuit Layout for the Electronic Organ

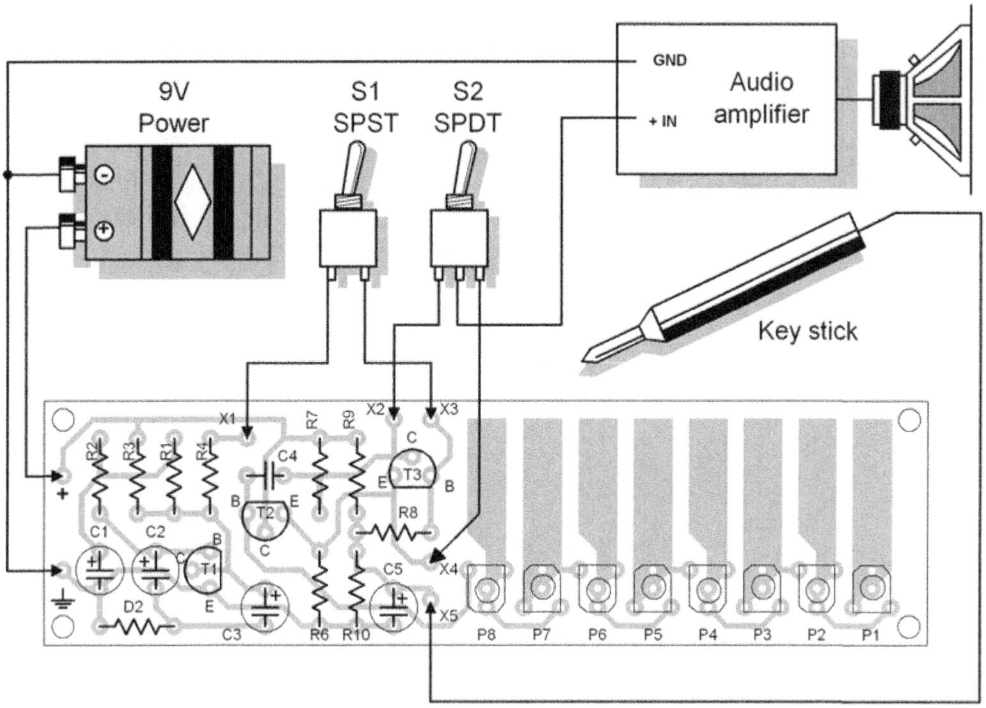

Figure 25.1 Parts Placement Layout for the Electronic Organ

26 SIGNAL CLIP INDICATOR

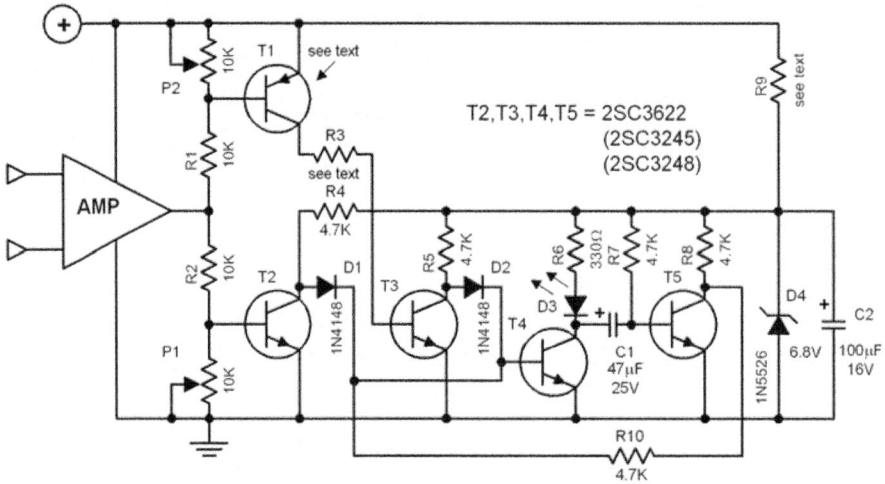

Diagram 26.0   Signal Clip Indicator

This circuit shows the clipping of an output signal coming out of a preamp or a final amplifier through a short lighting of a LED. The supply voltage of the circuit is not critical - it can be either symmetrical or asymmetrical. This circuit is normally connected as part of the amplifier circuit and uses, therefore, the existing supply lines of the amplifier. The monitored output signal is sampled at the point before the electrolytic output coupling capacitor as shown in the diagram 26.1. The LED lights up at clipping levels in both positive and negative swings of the signal.

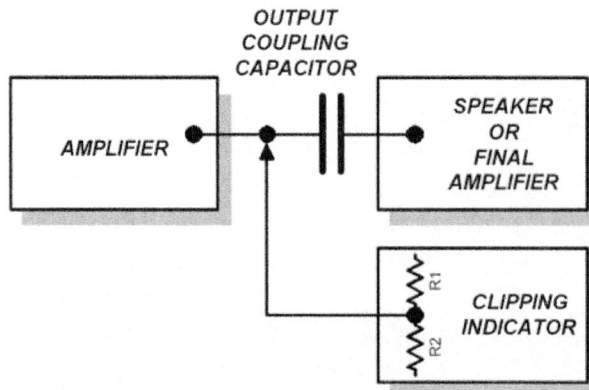

Diagram 26.1   Block Diagram

E C B

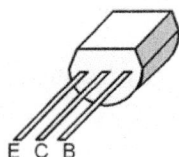

2SC3622   2SA970
2SC3245   2SA1136
2SC3248   2SA1137

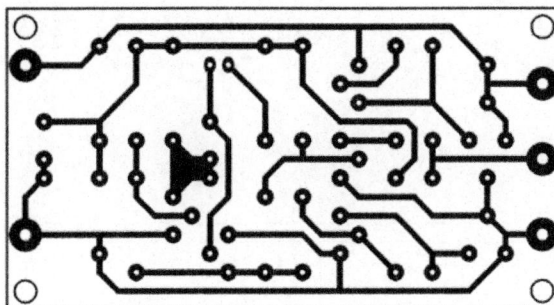

Figure 26.0   Printed Circuit Layout

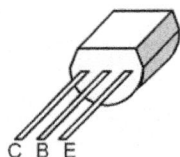

C B E

2SC1285
2SC1285A

Figure 26.1   Parts Placement Layout

The transistor T1 must be selected according to the supply voltage level. If the supply voltage is lower than 40 Volts, use one of the ff: transistor types: 2SA970, 2SA1136, 2SA1137. If the supply voltage is between 40 and 65 Volts, use either 2SC1285 or 2SC1285A. Select the value of R4 to let a current of around 12 mA flow through it.

It is also important to maintain a constant threshold level for the monoflop. To do this, select the value of resistor R9 so that the current flowing through the zener diode D4 is between 20...24 mA

The calibration of the circuit can be done quickly and accurately by using an oscilloscope. To calibrate: First, connect the oscilloscope at the junction of resistors R1 and R2. Then, inject a positive signal strong enough to cause clipping of the positive peak and adjust P2 until the LED lights up. Conversely, inject a negative signal and adjust P1 until the LED lights up.

27 GUITAR SOUND EFFECT

+12V
+
C3
1μF
25V

IN
C1
0.47

P1
25K

2
IC1
741
3
+
7
6
4

R1
2.7M

R2
1.2K

2
IC2
741
3
+
7
6
4

D1
D3
D2
D4
D1...D4 = 1N4148

P2
100K

C2
0.47

A
B
S1
OUT

-12V
-
C4
1μF
25V

Diagram 27.0 Guitar Sound Effect

This sound effect circuit is actually a frequency doubler. It is commonly used by rock guitarists. It is also called octave shifter. It is one of the standards in a guitarist's arsenal of sound synthesizers and special effect devices. The guitar tone coming from the amplifier is shifted by this circuit by one octave.

Potentiometer P1 sets the input level of the signal and P2 sets the output amplitude of the amplified signal. The gain of the IC must be set through P1 to the point just short before the signal clips.

Diodes D1 to D4 function as a bridge rectifier. These diodes also double the signal frequency. Both diodes are connected in feedback to the amplifier so that their nonlinear character cannot affect the signal. Switch S2 is a bypass switch to turn off the circuit once a normal guitar tone is desired. This frequency doubler not only doubles the frequency of the signal but also changes its form. The output tone sounds synthesized compared to the original.

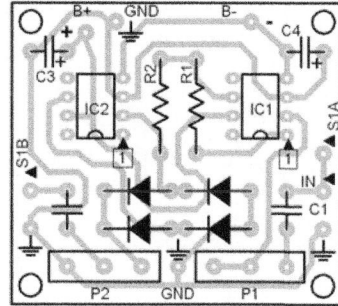

Figure 27.0   Printed Circuit Layout

Figure 27.1   Parts Placement Layout

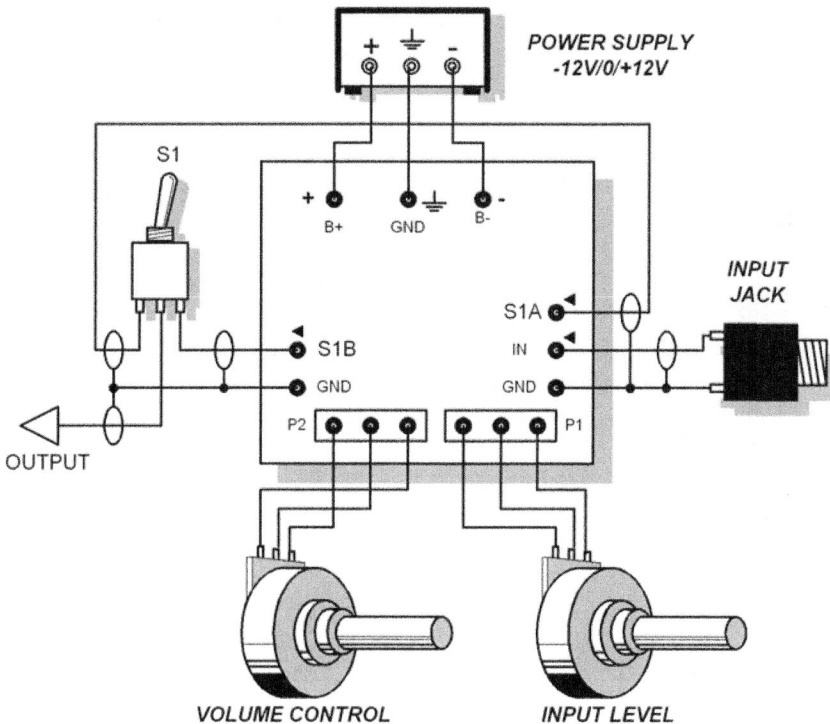

Figure 27.2   External Wiring Layout

28 HIFI HEADPHONE AMPLIFIER

Diagram 28.0 HIFI Headphone Amplifier

This amplifier is normally used to drive a headphone with a relatively low impedance. It provides 1 watt power output. It can also be applied as an output stage for preamplifiers in conjunction with active loudspeaker boxes. The circuit is composed of an op-amp and an additional transistor amplifier. Input signals pass through a low pass filter composed of R1-C2. Its application together with a relatively "fast" op-amp provides a low distortion factor. The standby current is preset by diodes D1...D4 and R7-R8. The feedback resistors R3 and R4 set the amplification factor at about 15 dB. The distortion factor is around 0.1 % with a bandwidth of 10 Hz to 30 kHz.

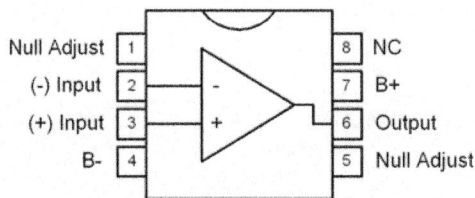

LF 356

	T1	T2
PAIR 1	One of the ff:	One of the ff:
	2SC3420	2SB1143
$U_c = 45$ V$_{max}$	2SD826	2N6414
$I_c = 1.5$ A$_{max}$	2SD1685	
	2N6412	

	T1	T2
PAIR 2	One of the ff:	One of the ff:
	2SD781	2SB874
$U_c = 100$ V$_{max}$	2SD1177	2SB874.B
$I_c = 1.5$ A$_{max}$	2SD1684	2SB874-C
	MJE243	2SB1144
	MJE244	MJE253

Table 28.0   Transistor equivalents for the transistor pair T1 & T2

The amplifier provides a maximum output of 1 watt to an 8 ohm load by an input level of 500 mV. High impedance headphones can be connected. The circuit can also drive a 4 ohm speaker.

In order to protect the output transistors from being destroyed in case of an output short circuit, they must be heatsinked. The transistors must also be electrically isolated from the heatsink. Supply voltage can be provided by an adaptor with 6-8 volts DC output.

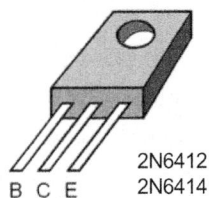

2N6412
2N6414

B C E

E C B

2SB874	2SD826
2SB1142	2SD1177
2SB1143	2SD1684
2SC2270	2SD1685
2SC3420	MJE243
2SD781	MJE244
	MJE253

Transistor equivalents

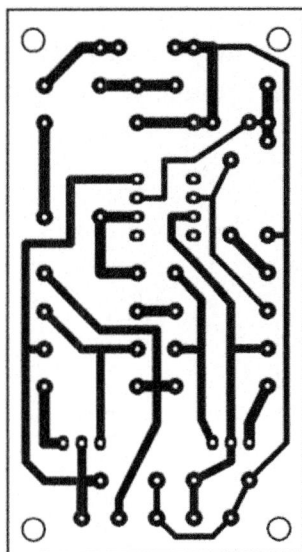

Figure 28.0
Printed Circuit Layout

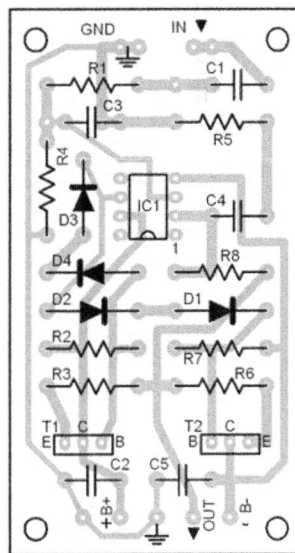

Figure 28.1
Parts Placement Layout

29 AUDIO FREQUENCY GENERATOR

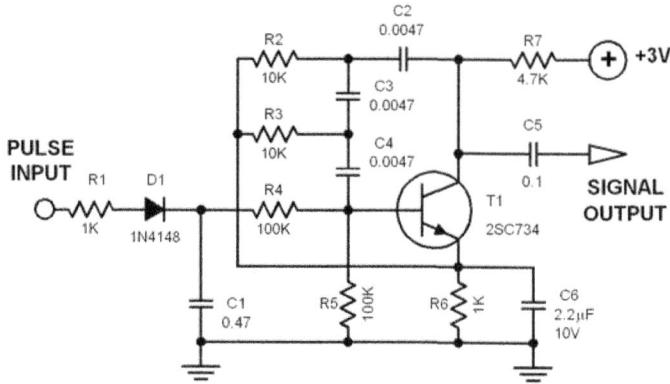

Diagram 29.0   Audio Frequency Generator

This electronic circuit is a triggered signal generator. When a positive pulse of about 6 volts (minimum) is fed to the circuit's input, a modulated audio signal comes out of the output. The signal pattern is similar to a bird's chirp. The pulse width of the trigger signal must be a minimum of 2.5 milliseconds. The voltage supply is between 9 volts and 20 volts. The circuit consumes about 2 mA or less. If the circuit is applied as a morse code signal generator, replace C1 with a 0.1 µF capacitor.

Figure 29.0
Printed Circuit Layout

Figure 29.1
Parts Placement Layout

30 SPEAKER PROTECTOR

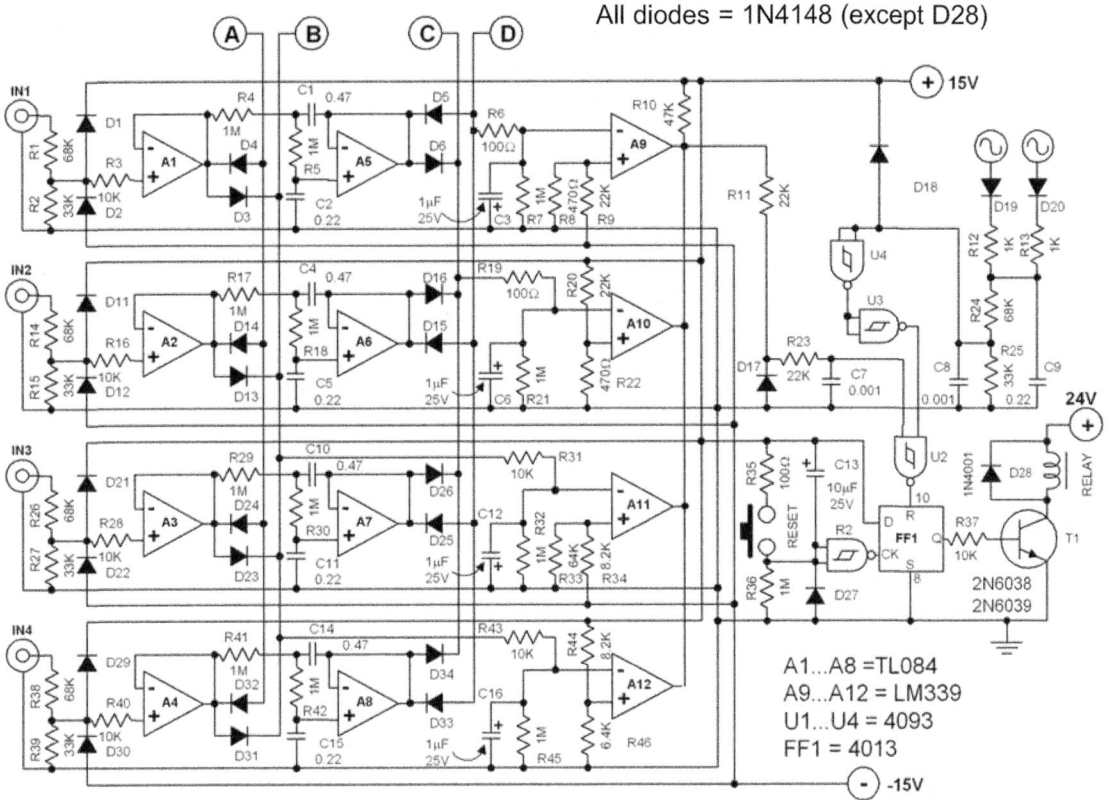

Diagram 30.0 Speaker Protector

This circuit protects the speaker when the final amplifier malfunctions. Conversely, it protects the final amplifier when the speaker overloads it. The protector circuit is originally designed for active-box builders who often encounter malfunctions in the modules they are experimenting.

The electronic circuit is composed of three stages: the input buffers, the lowpass filter and the relay stage. The input stage can monitor four signals. The number of inputs can be readily expanded according to actual needs. The relay will activate when a high input signal comes out of the final amplifier or when the speaker overloads the final stages. The speaker is connected to the contact terminals of the relay so that the speaker will be automatically disconnected when a malfunction occurs. S1 resets the relay.

31   TWEETER GUARDIAN

Small but very useful. This circuit provides protection to your tweeter specially when it is always driven near the maximum level. The first circuit uses a simple bulb as a load. This bulb will glow when the signal going to the tweeter reaches a certain preset threshold level.

The bulb functions like a positive temperature coefficient (PTC) resistor - meaning its resistance increases proportionally to its temperature. At around 5.5, the transistor 2N3055 will conduct and shorts the tweeter's input line to ground, thereby preventing the overloading of the tweeter.

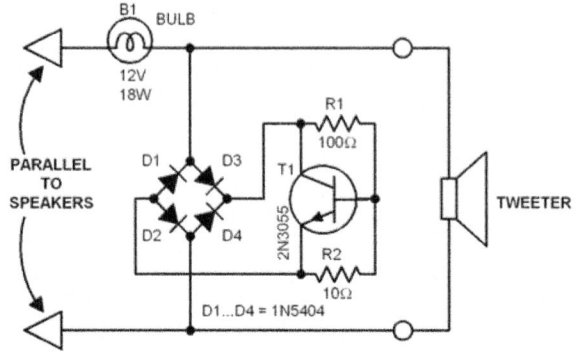

Diagram 31.0   Tweeter Guardian

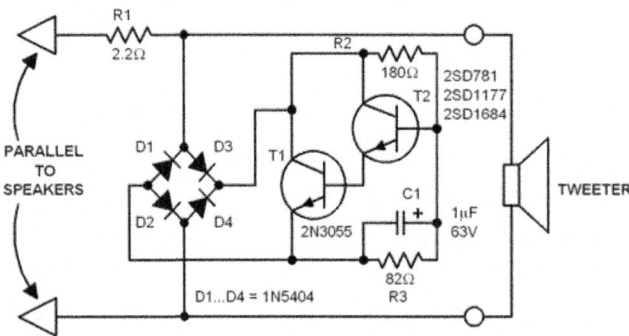

Diagram 31.1   Tweeter Guardian
(improved version)

The second circuit is an improved version. A fixed resistor replaces the bulb and the switch transistor is a darlington pair with a delay capacitor. The circuit still functions like the first circuit. However, due to the capacitor, the circuit will not react to a single overload peak.

This electronic circuit is usually installed inside high power boxes like those in disco applications where equipments are commonly driven to their maximum ratings.

E C B

2SD781
2SD1177
2SD1684

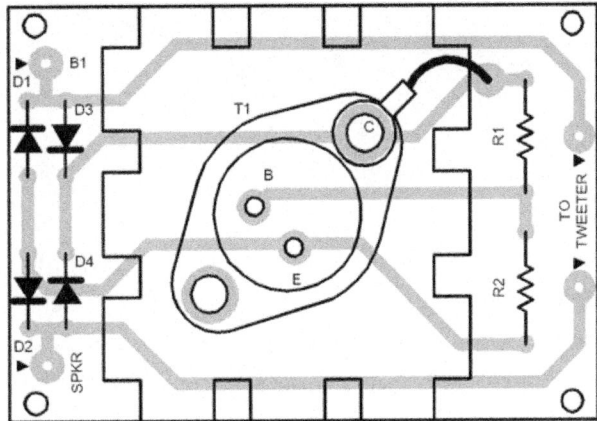

Figure 31.0 Printed Circuit Layout

C

B E

2N3055
Bottom view

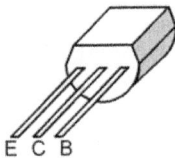

Figure 31.1 Parts Placement Layout

E C B

2SA970   2SC3622
2SA1136   2SC3245
2SA1137   2SC3248

Parts List for improved version

T1 = 2N3055
T2=2SD781 (2SD1177)(2SD1684)

D1...D4 = 1N5404

R1 = 2.2 Ω
R2 = 180 Ω
R3 = 82 Ω
C1 = 1 µF/63 volts

32   AUDIO MIXER

Diagram 32.0   Audio Mixer (emitter-driven-single transistor)

This single transistor mixer circuit is intended for audio applications and is designed to be as tolerant as possible in the type of the transistor used. Additionally, the number of input lines is flexible. You simply need to change the value of a single resistor to adapt the circuit to the actual number of inputs. The type of the transistor determines the values of the biasing components

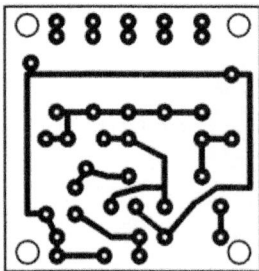

Figure 32.0
Printed Circuit Layout

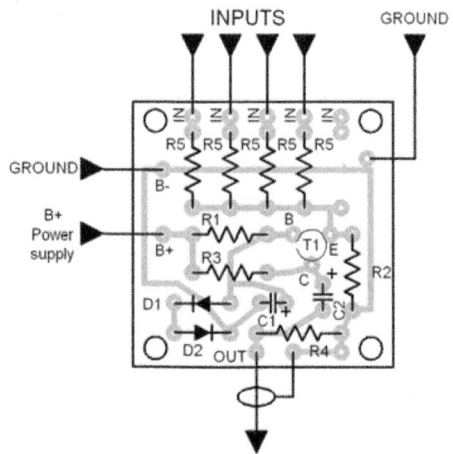

Figure 32.1
Parts Placement Layout

The resulting output voltage is dependent on this formula:

To bias the circuit to other supply voltage some values must be changed according to the following formulas: The collector current of the applied transistor can be computed using this formula:

where i = the collector current.

To drive the circuit to maximum, the value of the resistor R3 must be computed using this formula:

After determining the value of R3, the value of R5 can be simply found by using this formula:

... where n represents the number of the actual inputs used. As can be seen in the circuit diagram, you can decide how many inputs you can connect as long as you follow the given formula to compute the value of resistor R5.

$$U_{output} = R3/R5 \; (U1 + U2 + U3 + ...)$$

$$i = \frac{0.6 \text{ volts}}{R2}$$

$$R3 = \frac{VCC}{1.2 \text{ volts}}$$

$$R5 = R3(n)$$

33   TOUCH VOLUME CONTROL

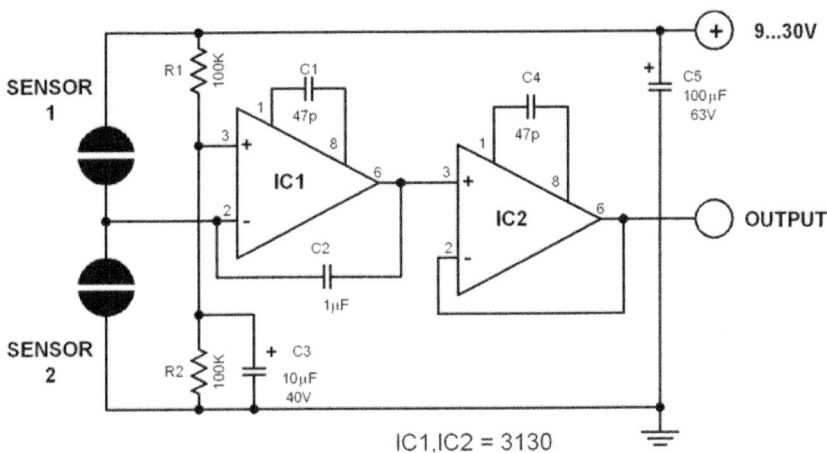

Diagram 33.0 Touch Volume Control

Touch controls are not only used to switch devices on or off. They can also be used to control different functions. One good example is the TV remote control. If it is very important to keep the activated functions for a long period of time, it is always better to use a digital memory system. However, if small drifts in the control status is acceptable, a simple analog design can be used to memorize the status.

The featured electronic circuit is one such analog memory touch control switch. The main function centers mostly on the IC1. It is an op-amp configured as an integrator with a high impedance input. If sensor 1 is touched, the capacitor C2 charges through the skin resistance and voltage at the output of IC1 decreases linearly until it reaches zero volt.

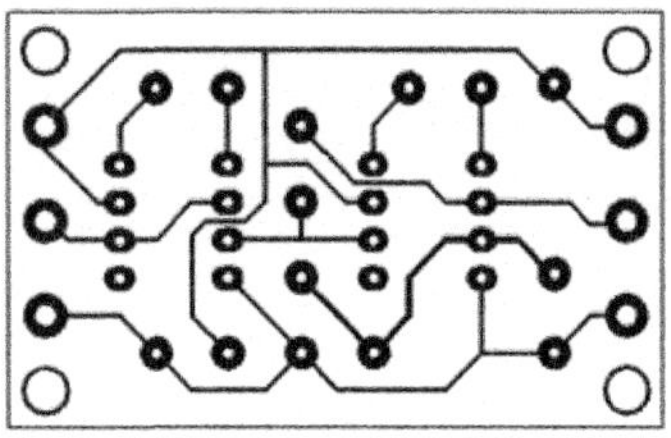

Figure 33.0
Printed Circuit Layout

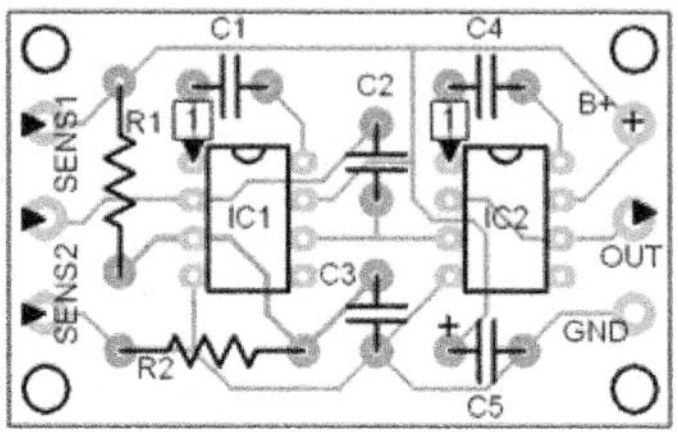

Figure 33.1
Parts Placement Layout

Touching the other sensor (sensor 2) will produce the opposite result: the voltage at the pin 6 of IC1 will increase linearly until it reaches the power supply level. The special function of this circuit is that after moving the finger away from the sensor(s), the output voltage of IC1 stays at that level. This voltage value is memorized by C2. This analog memory however has a problem in long time periods: the voltage value drifts away by 2% per hour due to the unavoidable current leak in the capacitor. To improve this situation, it is highly recommended to put the circuit in a moisture proof box.

This electronic circuit has a wide application range. It can be used in devices where a potentiometer can be controlled with variable voltage levels. The touch sensors can also be replaced with conventional push button switches. The capacitors C1 and C4 are very important in the circuit: they prevent the IC1 from oscillating. Simultaneously closing both switches will not damage the circuit.

34 AUDIO WATTMETER

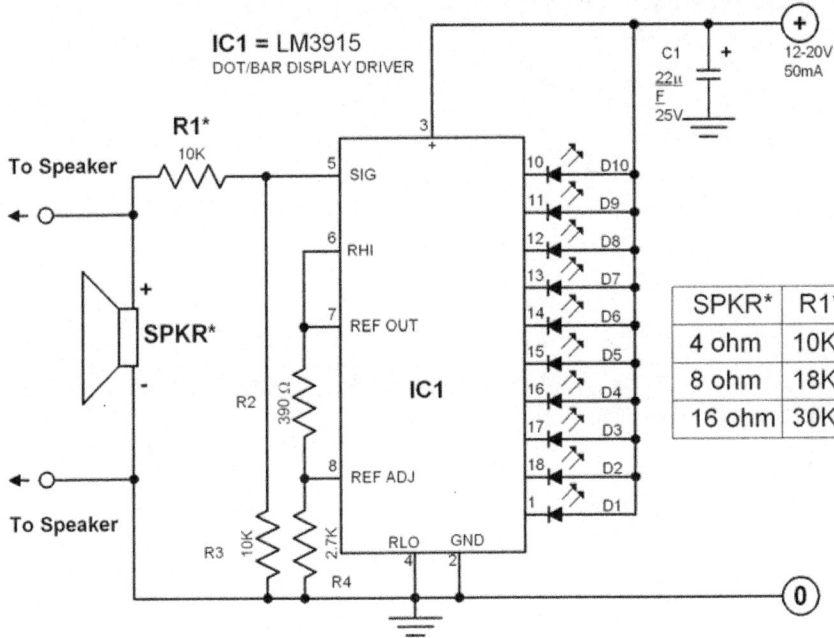

IC1 = LM3915
DOT/BAR DISPLAY DRIVER

SPKR*	R1*
4 ohm	10K
8 ohm	18K
16 ohm	30K

Diagram 34.0 Audio Wattmeter

You want to measure the power output of your audio amplifier? Well, take one LM3915 IC and add a few passive components .. then viola! You've got a simple but effective audio wattmeter.

This electronic circuit uses a row of colored LEDs as a scale to show the relative power output of your amplifier in watts. It can be easily inserted into the speaker box. All you need to do is hook a supply voltage to it. The value of R1 depends on the impedance of the speaker being used. The small table in the diagram shows the necessary values of R1.

If you want to apply the circuit to stereo systems you must build two identical circuits.

The supply voltage of the circuit is a 12...20 volts/50 mA DC adapter. Take note that the LM3915 only displays the positive swing of the signal. In testing amplifiers it does not matter anyhow since a sine wave is normally used as a test signal. To test the circuit without using an actual speaker, you can connect a dummy resistor with a value of either 8 or 4 ohms.

Figure 34.0
Printed Circuit Layout for
the Audio Wattmeter

Figure 34.1
Parts Placement Layout for
the Audio Wattmeter

35 LOW NOISE PREAMP

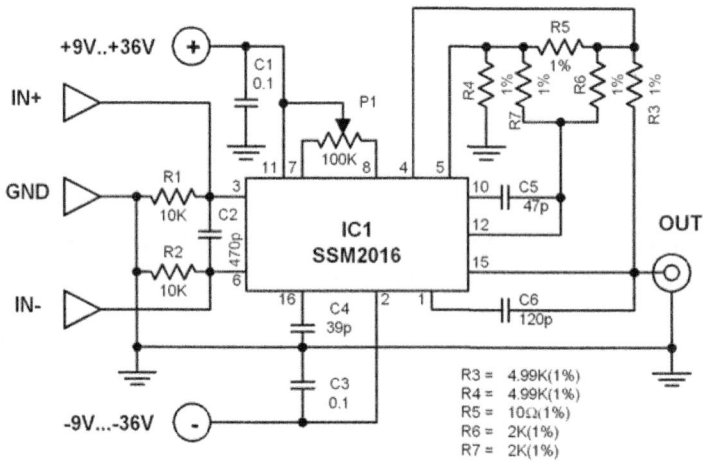

Diagram 35.0 Low Noise Preamp

R3 = 4.99K(1%)
R4 = 4.99K(1%)
R5 = 10Ω(1%)
R6 = 2K(1%)
R7 = 2K(1%)

This is a single IC low noise preamp made possible by using the SSM2016 from PMI. It has a symmetrical input and extremely low noise characteristics. It is highly capable of amplifying signal sources with low internal resistance (less than 1 kiloohms) like moving coil microphones and dynamic microphones with impedance ranging from 10 ohms to 600 ohms.

The amplification is adjustable by changing the resistance of R5 with a value between 3.5 and 1000 ohms. The amplification *A* can be found with the ff: formula:

$$A = (R3+R4)/R5+(R3+R4)/(R6+R79)$$

It can be simplified to A= 10K/R+3.5 using the given component values. If for example R5 = 10 ohms, then the amplification is 1000. The offset voltage at the inputs can be set to zero through potentiometer P1. It is highly recommended to use metal film resistors in constructing the circuit.

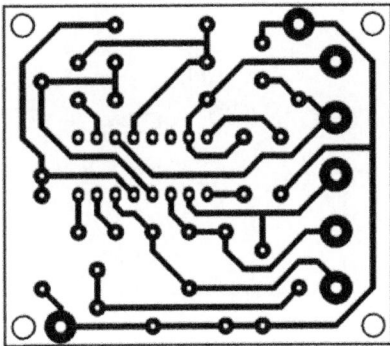

Figure 35.0   Printed Circuit Layout

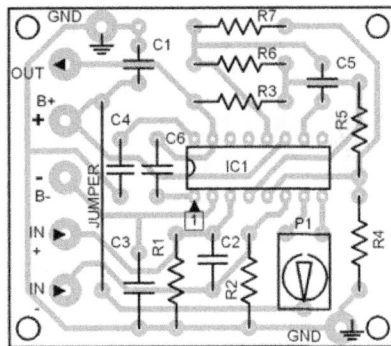

Figure 35.1   Parts Placement Layout

Figure 35.2   External Wiring Layout

36 HEATSINK THERMOMETER

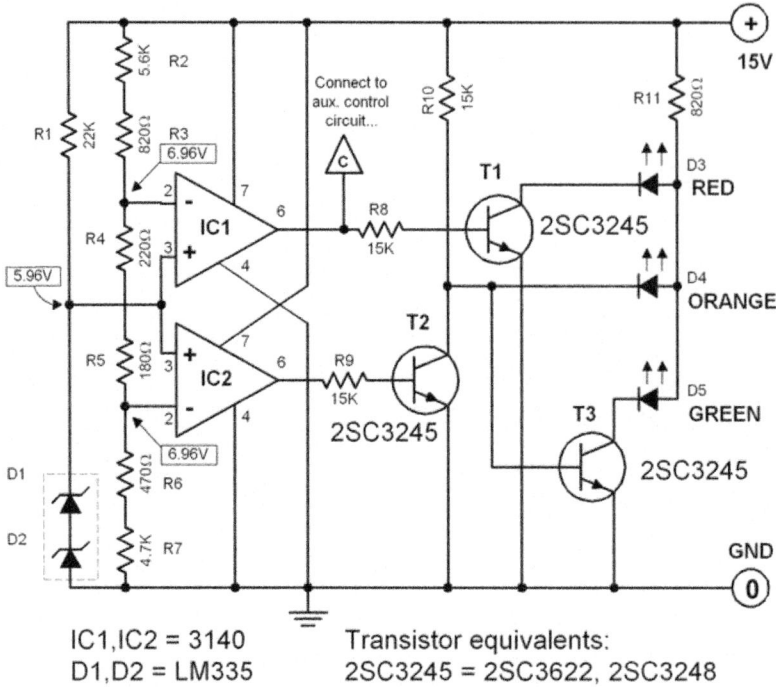

Diagram 36.0   Heatsink Thermometer

This "thermometer" monitors the temperature of a heatsink and displays its approximate level through three LEDs. The green LED lights up as long as the temperature is less than 50°C. The orange LED lights up when the temperature is between 50°C and 75°C. The red LED lights up when the temperature rises above 75°C.

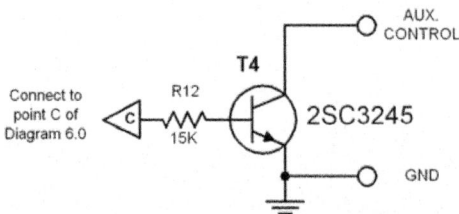

Diagram 36.1 Aux control driver circuit

The temperature sensor is made of two special zener diodes. The zener voltage is exactly 5.96 V when the temperature is 25°C. This zener voltage increases by 20 mV per degree centigrade.

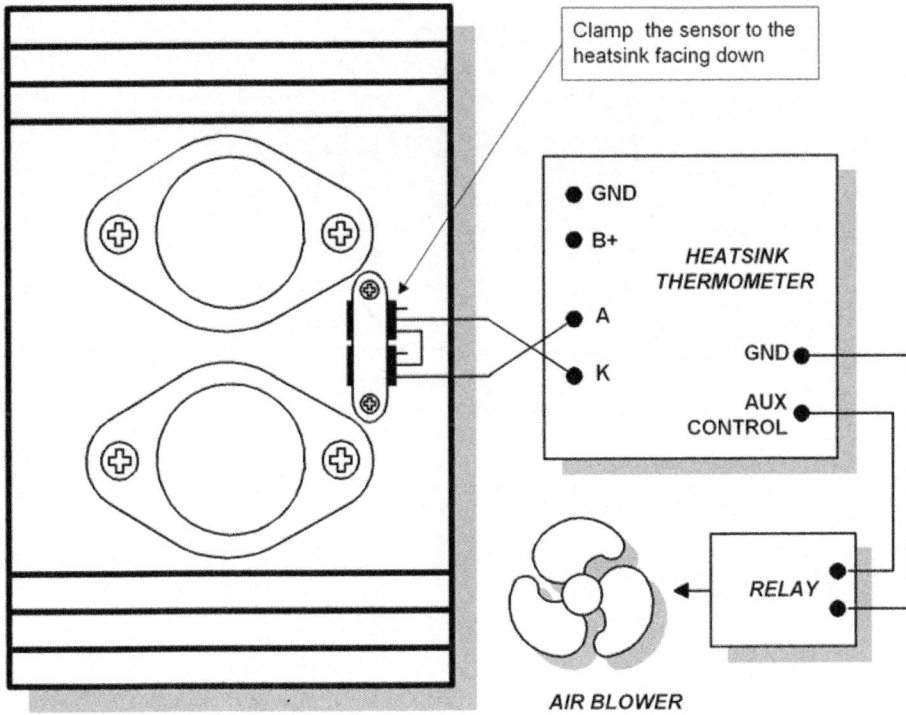

Figure 36.0 External Wiring Layout

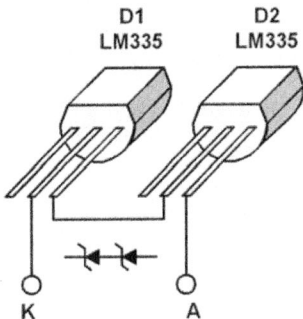

Sensors D1 and D2 must be installed on the heat sink as close as possible to the final transistors to minimize thermal losses.

Of course, you can use this temperature sensor for purposes other than blinking LEDs. One of the best applications of this circuit is the automatic control of an air blower that cools the heatsink down to a safe temperature level. Connect a relay to transistor T4. The relay, in turn, controls any kind of device connected to it like, for example an air blower. Never connect the device directly to transistor T4.

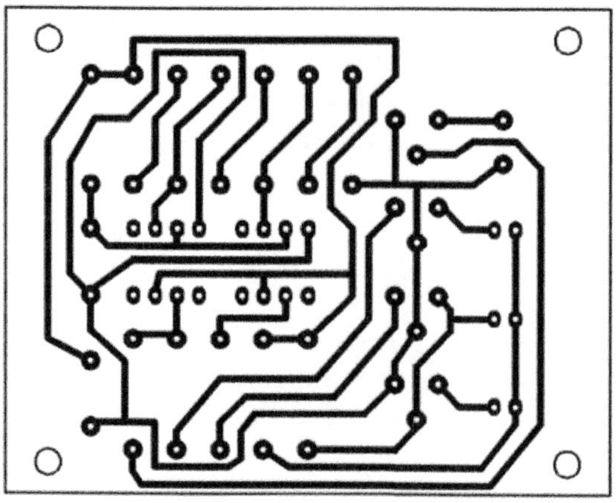

Figure 36.1 Printed Circuit Layout for
Heatsink Monitor

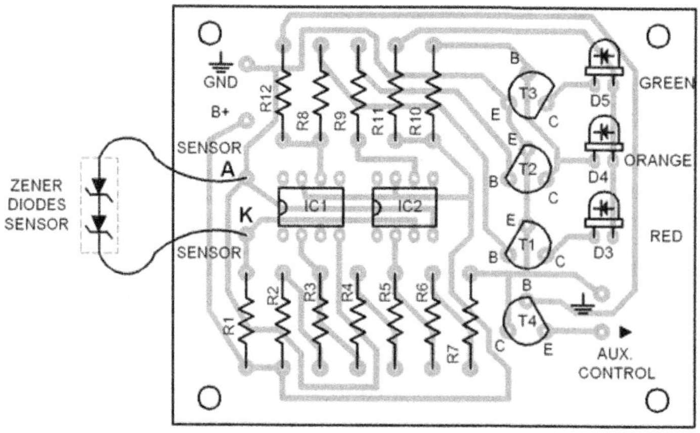

Figure 36.2 Parts Placement Layout for
Heatsink Monitor

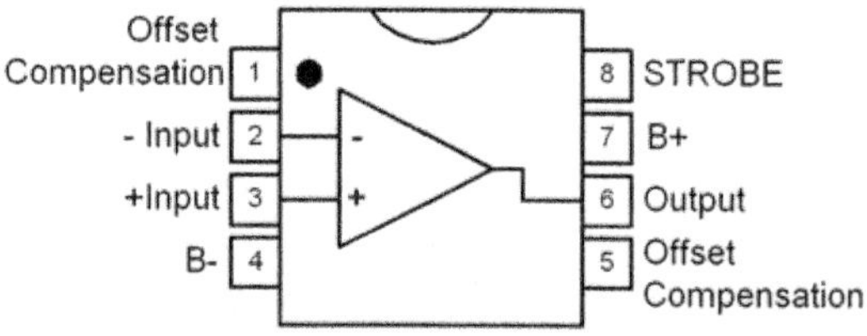

CA3140
BIMOS OPAMP

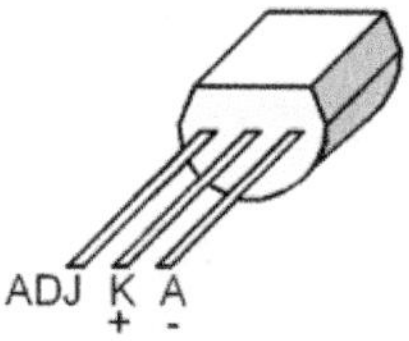

LM335

37   3- WAY AUDIO EQUALIZER

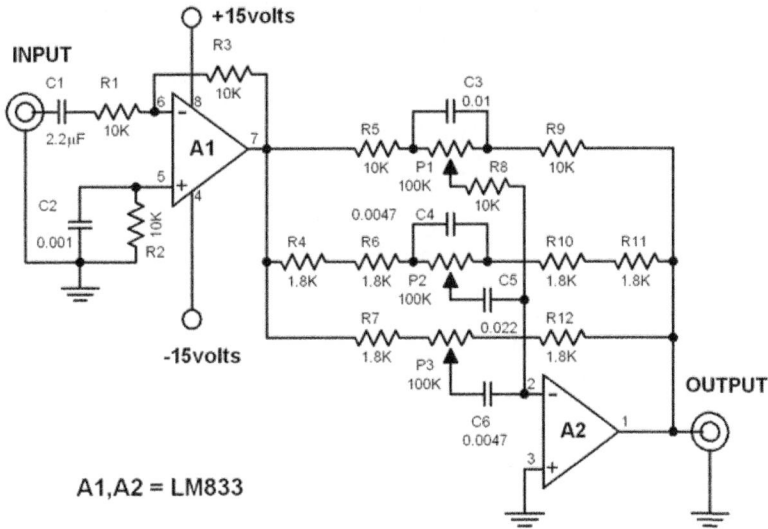

Diagram 37.0   3-Way Audio Equalizer

The circuit is an active filter network for bass, mid and high audio ranges. It is designed around the LM833, an op-amp from National Semiconductors. This op-amp has the following characteristics: very low noise figure (4.5 nV/sqr(Hz)), wide bandwidth (15 MHz at Vu = 1), and relatively high slew rate (7 V/μs). The output is designed to be dc-coupled, however due to slight dc variations through the 100K potentiometers at the feedback lines of the opamp A2, a coupling capacitor might be needed.

Technical specifications:

The cutoff frequencies:

> bass range = 200 Hz
> high range = 2 kHz

The midrange is a bandpass network with a center frequency of 1 kHz. The maximum equalizer range is about 16 dB. In the middle position of potentiometer, the noise attenuation is about 90 dB with a banwidth of 1 MHz and a gain of 0 dB (zero amplification). The gain can be changed through R2 using the following formula:

Vu = R2/R1.

38   AUDIO PEAK METER

Diagram 38.0   Audio Peak Meter

This circuit can measure the peak level of AC signals regardless of their waveform. It is independent from the signal direction. Both positive and negative going peaks deliver the same result. As you can see in the block diagram, block A is a rectifier for the positive going peaks while block B is a rectifier for the negative going peaks. The rectified peaks are then inverted in block C and added to the rectified signal coming from block A.

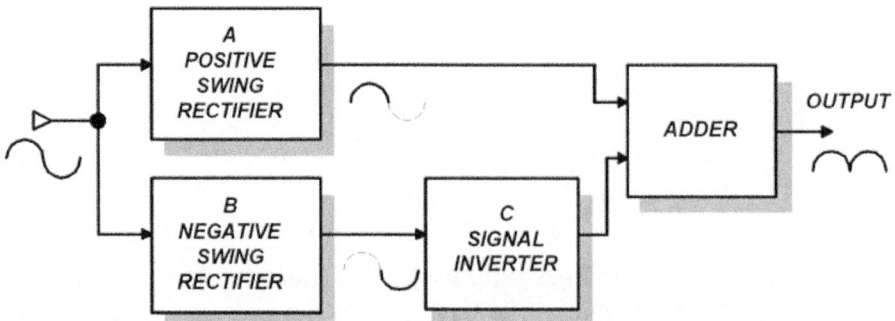

Diagram 38.1   Block Diagram

A voltage of around 600 mV appears at the circuit's point C. The highest peak value appears at the resistor R16 of the output terminal. A meter can be connected at this point to display the measured value.

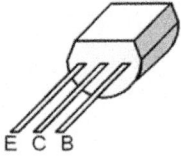

E C B

Bottom view

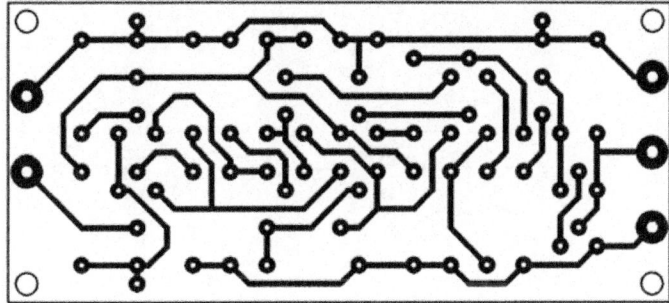

Figure 38.0 Printed Circuit Layout

2SC3622
2SC3245
2SC3248
2SA970
2SA1136
2SA1137
2SA1016
2SA1123

Transistor equivalents

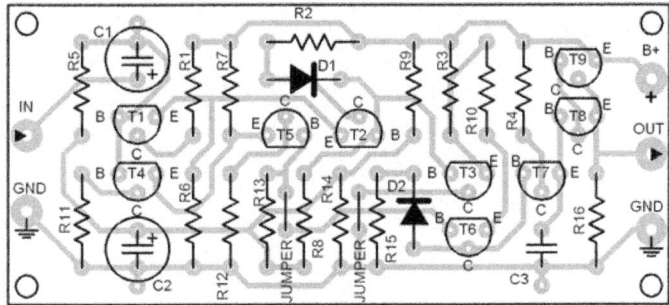

Figure 38.1 Parts Placement Layout

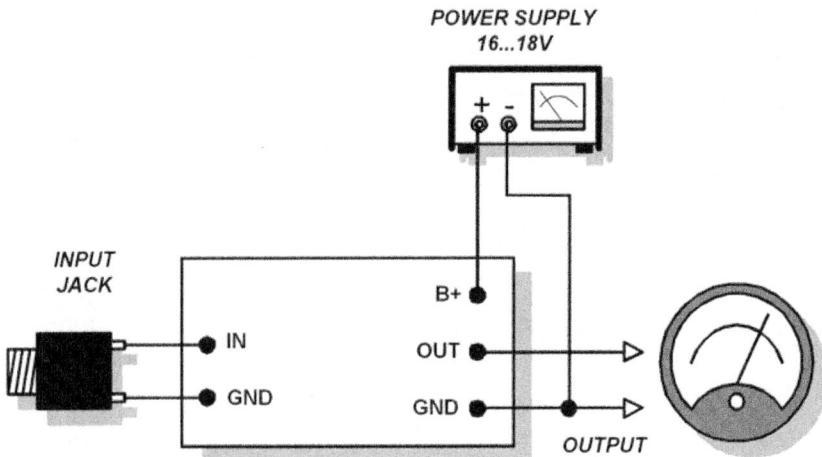

Figure 38.2 External Wiring Layout

39   NOISE FILTER

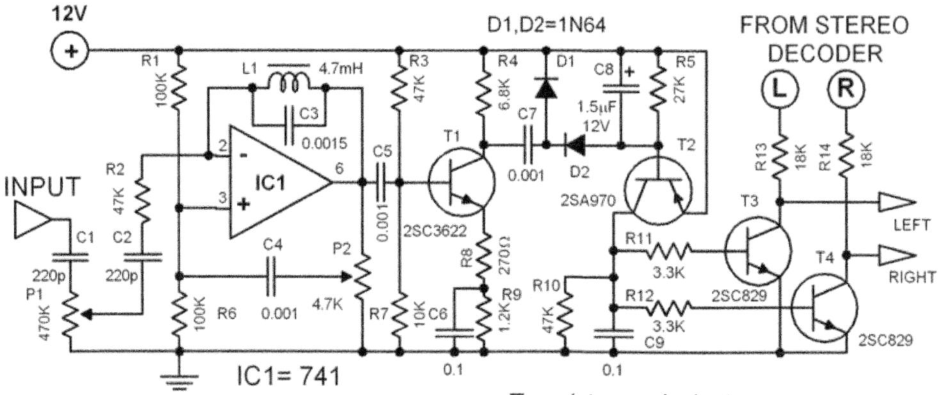

12V

D1,D2=1N64

FROM STEREO DECODER

INPUT

IC1= 741

Transistor equivalents:
2SA970 = 2SA1136, 2SA1137
2SC829 = 2SC460
2SC3622 = 2SC3245, 2SC3248

Diagram 39.0   Noise Filter

This circuit blocks the audio signal from going to the final amplifier when the noise level exceeds a certain preset value. The noise threshold level can be set through potentiometer P1. This noise filter works with the "squelch" method - it prevents the amplification of noise signals by preventing them from reaching the amplifier. One practical application of the circuit is to mute the audio amplifier of an FM receiver when an empty frequency is found during channel changing or searching (the noise level is very high in an empty channel or frequency).

Installing this electronic circuit is very easy. The noise filter is connected parallel to the demodulator circuit of the FM receiver (see Diagram 39.1). The input point C1 is connected to the demodulator output of the stereo receiver. The potentiometer P2 sets the selectivity of the circuit. P1 sets the sensitivity level. Transistors T2 and T3 short circuit the output signal from the stereo decoder, preventing it from reaching the audio amplifier.

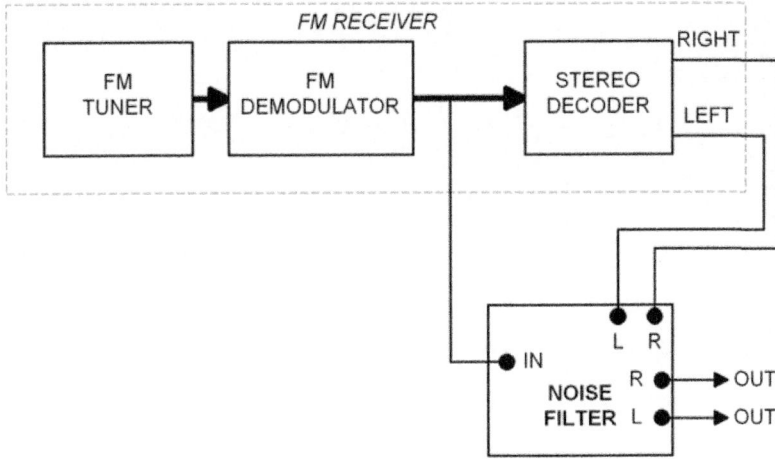

Diagram 39.1 Installation of Noise Filter

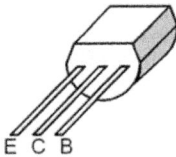

Diagram 39.2 Block Diagram of Noise Filter

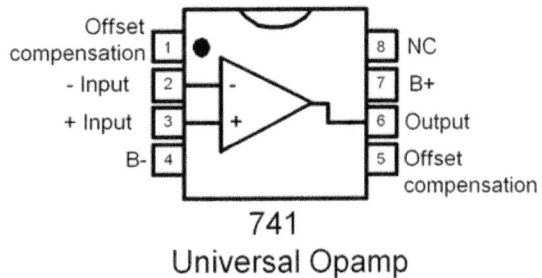

2SA970	2SC3248
2SA1136	2SC829
2SA1137	2SC460
2SC3622	2SC3245

Transistor equivalents

741
Universal Opamp

40 MICROPHONE PREAMP

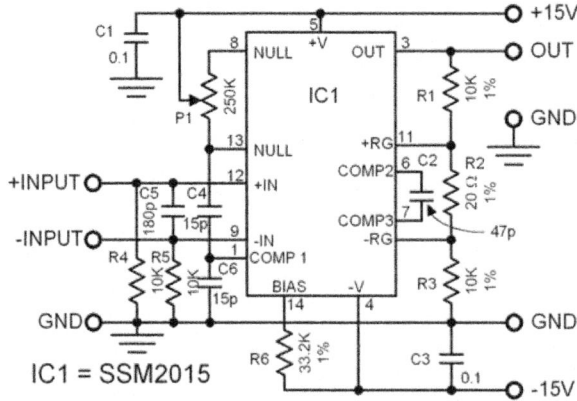

Diagram 40.0 Microphone Preamp

This microphone preamp is extremely quite exhibiting an extremely low noise factor of 1.3 nV/sqr(Hz). Additionally, it has a high voltage amplification factor between 10 and 2000. This factor (G) comes from the simplified formula:

$$G= (2000/R2)+3.5$$

when R1 = R3 = 10 KΩ

In the featured application circuit, the amplification factor is set to 1000. Resistor R6 affects the bandwidth and slew rate. The value of 33k for R6 provides a good compromise. This value, however, causes a standby current of 4.5µA. Anyway, we can achieve a noise factor of 95 dB with the inputs short circuited or 86 dB with an input impedance of 600 ohms. If the value of R6 is changed, the value of C4 and C6 must also be changed. Table 40.0 shows the necessary values.

Table 40.0

R6	C4	C6
27K...47K	15p	15p
47K...68K	10p	15p
68K...150K	5p	20p

Table 40.1

G	R6 = 27K..47K	47..68K	68K..150K
10	500K	250K	250K
100	500K	100K	100K
100	250K	100K	50K

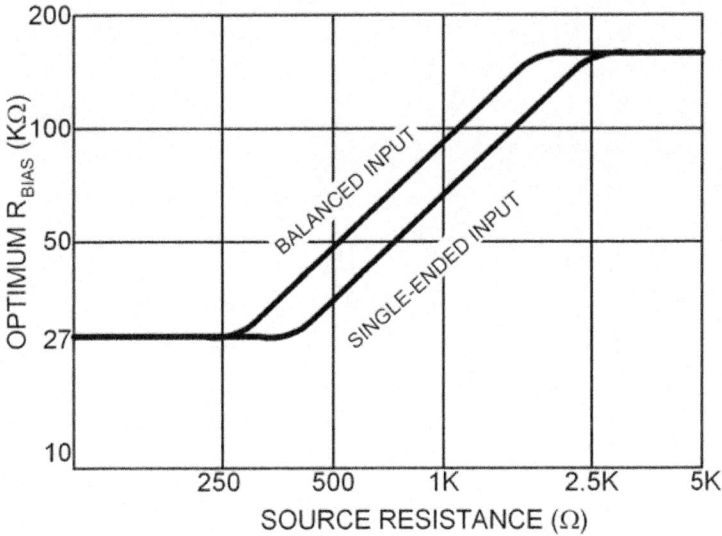

Graph 40.0 Source to Bias Resistance

The differential inputs of SSM2015 "float" that is why resistors R1 and R2 are added. These two resistors stabilize the DC current input. If the chip is used for so called single-ended applications, it is necessary to adjust P1 to compensate for input offsets. See Table 40.1.

Potentiometer P1 does not affect the amplification factor of the IC. The IC can drive capacitive loads up to 150pF. It has a bandwidth of 180 kHz (-3dB). The output voltage is 3 volts with a load of 1 kiloohm.

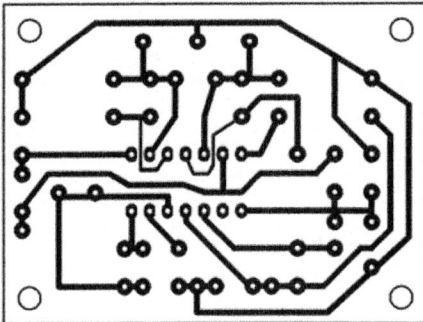

Figure 40.0 Printed Circuit Layout

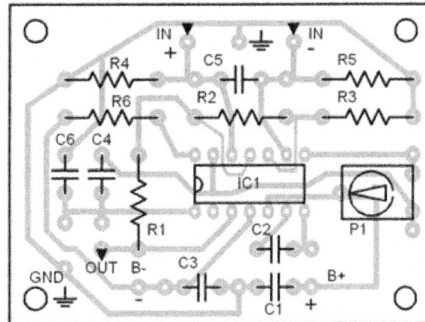

Figure 40.1 Parts Placement Layout

41 MUSIC PROCESSOR

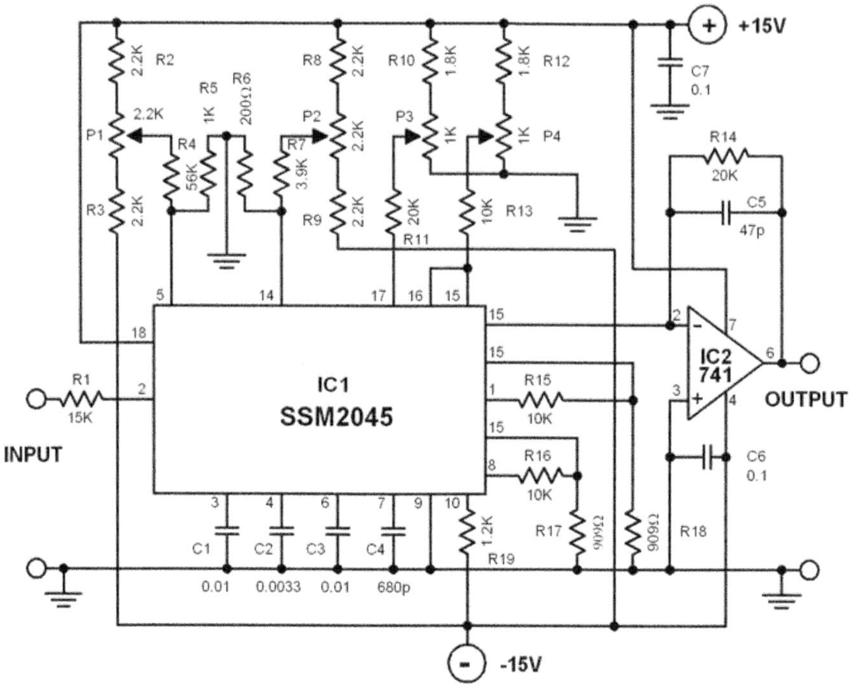

Diagram 41.0 Music Processor

This electronic circuit features the SSM2045 IC which was developed specially for electronic music applications. The IC features a universal application. The circuit is basically configured as a low pass filter with a DC voltage control for gain. The input signal is set to a working level of 150mVp-p through the resistor R1.

The filter has two buffered outputs: the 2-pole output at pin 1 and 4-pole output at pin 8. Internally, these outputs are connected to two voltage-controlled-amplifiers (VCA). The resistors R15 and R16 are connected to these outputs to achieve optimum offset and control voltage suppression. Potentiometer P4 is the volume control. The current that flows to the pins 15 and 16 should not go beyond the maximum of 250 µA. The balance of the two VCAs and the entire filter is being controlled by a voltage range of -250mV to +250mV at pin 14. This voltage can be set by potentiometer P2.

The input can be driven with source impedance up to a maximum of 200 ohms. With an input level of 0 dBm, the VCA weakens by 6 dB. The bias current needed at pin 17 (Q input) is between 120 µA and 185 µA. The cutoff frequency can be shifted between 20 Hz and 20 kHz with a variable voltage at pin 5. This can be varied though the potentiometer P1. The capacitor values were selected to give the filter its Butterworth characteristics.

The output current of the SSM2045 IC is converted to a voltage output by the 741 op-amp (IC2). Any subsequent circuit must be DC decoupled from IC2.

The noise-voltage ratio is about 80 dB.

42 SPEECH PROCESSOR

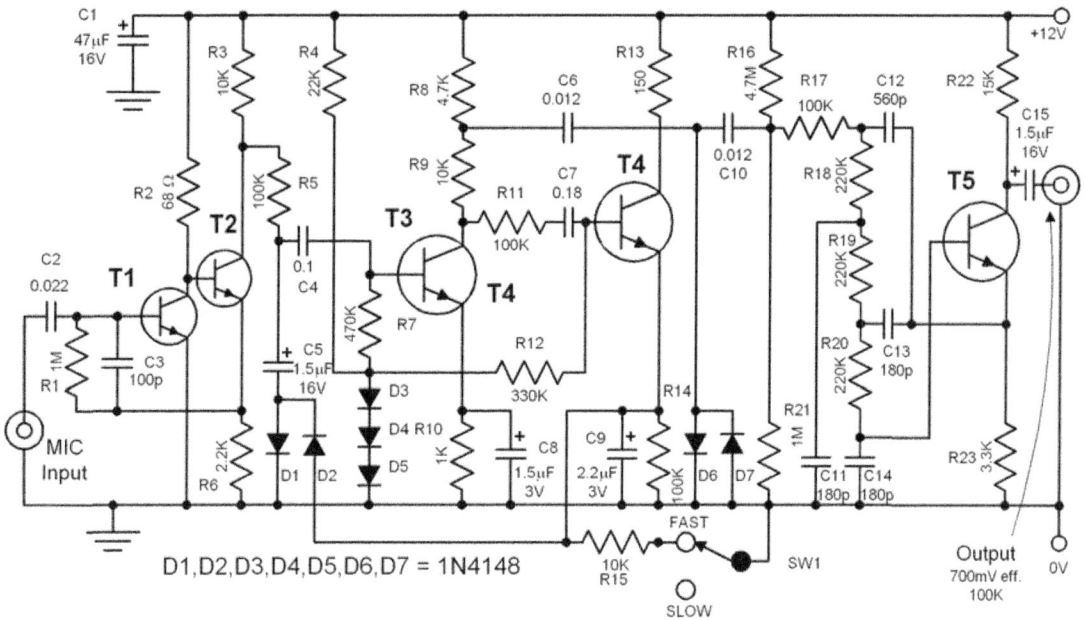

T1 = 2SC3112 / 2SC2675   T3,T4,T5 = can be one of the ff:
T2 = 2SC3622   2SC1876 / 2SD406 / 2SD614 / 2SD688 / 2SD892A /
                                                    2SD1153 / 2SD2116 / 2SD2117 / 2SD2213

D1,D2,D3,D4,D5,D6,D7 = 1N4148

Figure 42.0 Speech Processor

There are two ways to modulate a transmitter to obtain the best possible modulation result: clipping (clipping off the amplitude peaks) and dynamic compression (generation of a constant medium powered signal carrier). Both methods have their own disadvantages. In clipping, there is no constant signal carrier, and the transmitter has a tendency to be overmodulated or undermodulated, resulting to degradation of the transmitted information. In dynamic compression, degradation also happens because of the occasional uncontrolled signal bursts due to the time constant of the compressor's regulator.

To avoid these problems, both modulation techniques are combined in one circuit. First, the modulating signal is compressed to achieve a relatively constant signal level, and then clipped to remove the unwanted amplitude peaks. The circuit first amplifies the audio signal before it is introduced into the main processing circuits. The gain of the amplifier stage (T1,T2) depends upon the impedance of the microphone being used.

This technique accommodates both low and high impedance microphones and still achieves the same signal level. Diodes D1 and D2 function together as voltage controlled signal processor. Transistor T4 on the other hand controls these two diodes.

When the voltage at the anode of D3 exceeds the threshold voltage by about 0.5 volts, the incoming signal will be automatically reduced through the C5-D1 combination. Very fast signal bursts which escape the compressor circuit are clipped by diodes D6-D7. The amount of clipping is determined by the combination R8-R9. This speech processor is highly useful as modulator stage for AM, FM, and SSB transmitters.

The low pass characteristics of this processor is designed for 80 meter band operation. If one desires a low pass characteristic with a different cut-off frequency, multiply the capacitors C11 up to C14 by factor A.

The multiplication factor A can be found by using the following formula. For example, if the desired cut-off frequency is 6kHz, the capacitors C11 to C14 must be replaced with half of the original values.

Formula for A

$$A = \frac{3}{f}$$

(f = cut-off frequency in kHz)

43 AF GENERATOR

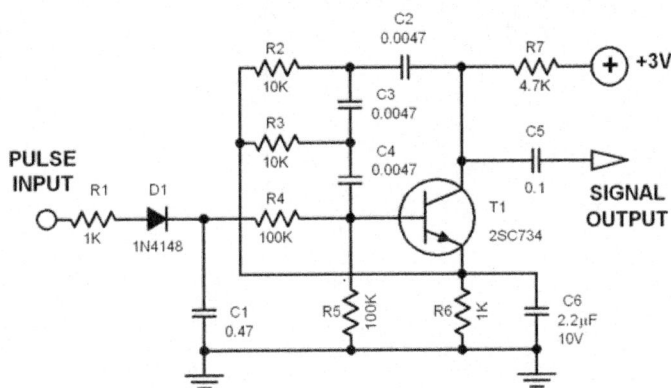

Diagram 43.0 AF Generator

This circuit is a triggered audio frequency (AF) signal generator. When a positive pulse of about 6 volts (minimum) is fed to the circuit's input, a modulated audio signal comes out of the output. The signal pattern is similar to a bird's chirp. The pulse width of the trigger signal must be a minimum of 2.5 milliseconds. The voltage supply is between 9 volts and 20 volts. The circuit consumes about 2 mA or less. If the circuit is applied as a morse code signal generator, replace C1 with a 0.1 µF capacitor.

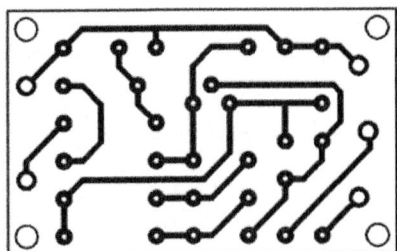

Figure 43.0
Printed Circuit Layout

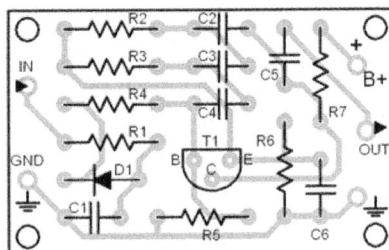

Figure 43.1
Parts Placement Layout

44 STEREO AUDIO MIXER

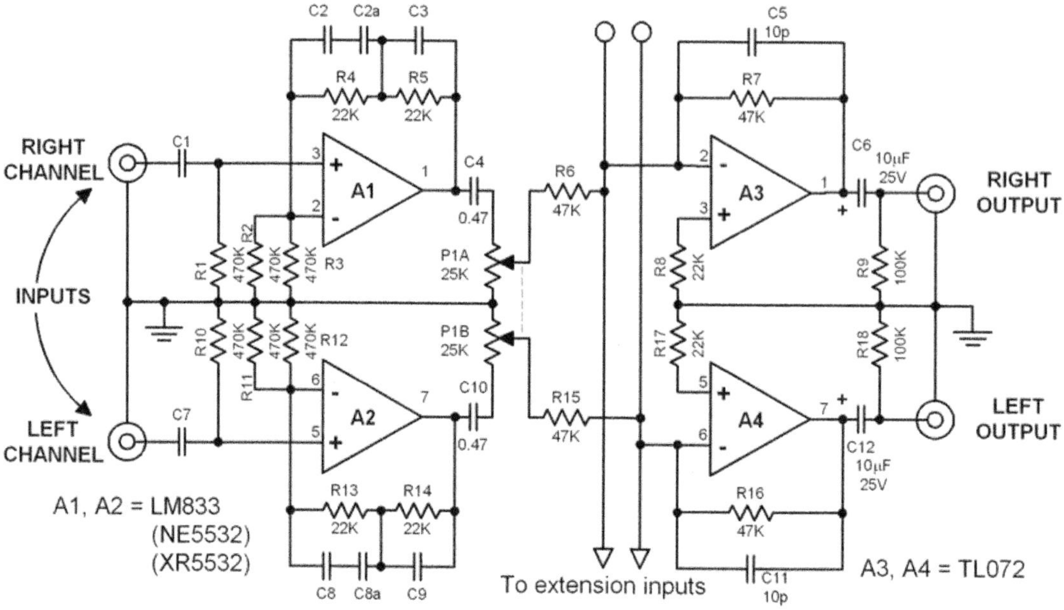

Diagram 44.0   Stereo Audio Mixer

This is a mixer circuit which can mix stereo sources. The component values for the input circuit varies according to the type of instrument connected to it. Table 44.0 shows the necessary component values for every input instrument. You will notice that if you design the input for a tape instrument you don't need the A1/A2 IC and the discrete components around it. You simply connect the source line to the capacitor C4 or C10.

input device	C1/C7	C2/C8	C2a/C8a	C3/C9	R1/R10	R2/R11	R3/R12	R4/R13	R5/R14
Mic-low ohm	10mF/10.2V	-	-	10p	680W	21K	-	short	100K
Mic - Hi ohm	0.47	-	-	10p	22K	1K	-	short	100K
Tape	remove A1 & A2, connect directly to C4 or C10								
Phono	0.22	0.0015	0.0015	0.0033	47K	2.2K	2.2K	100K	1M

Table 44.0

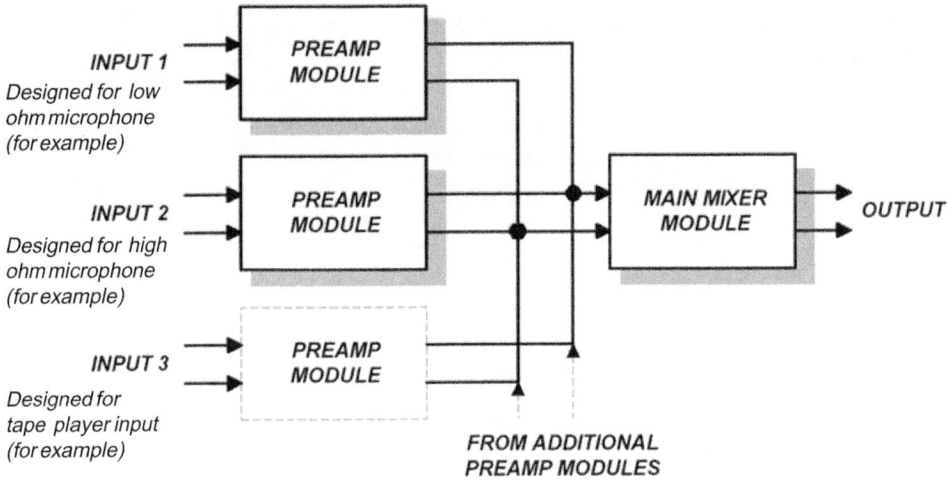

Diagram 44.1 Installation Diagram

Figure 44.0 Printed Circuit Layout of
the Input Preamp Module

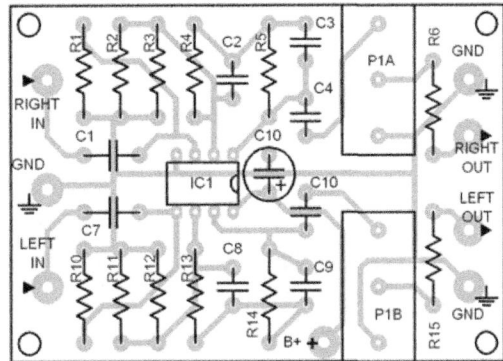

Figure 44.1 Parts Placement Layout of
the Input Preamp Module

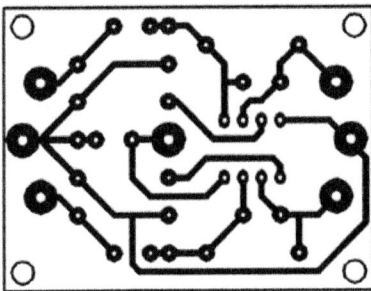

Figure 44.2 Printed Circuit Layout of
the Main Mixer Module

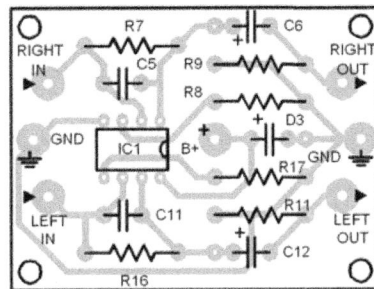

Figure 44.3 Parts Placement Layout of
the Main Mixer Module

45 STEREO INDICATOR

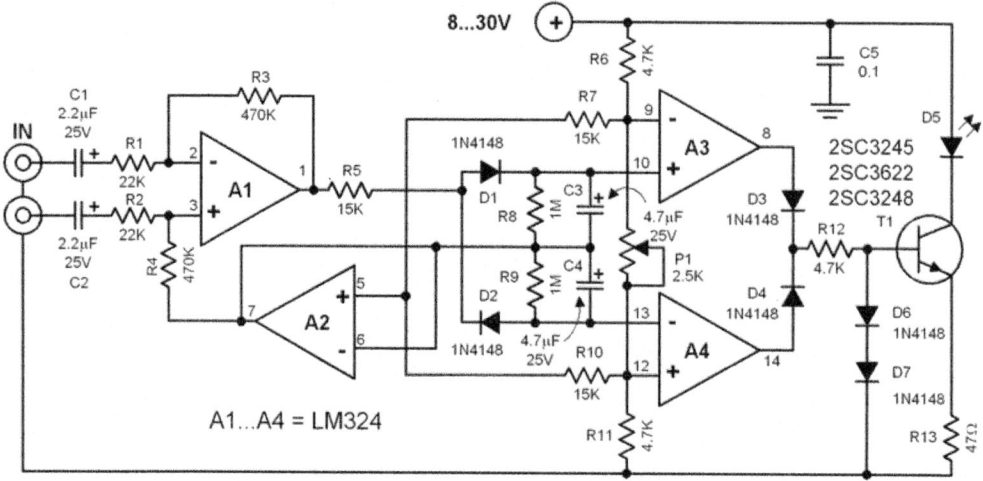

Diagram 45.0 Stereo Indicator

   This electronic circuit indicates whether the signal being received by your FM radio is a true stereo signal. The principle being applied is very simple: the left and right channels are compared. When a difference between the two channels appears, the signal is stereo. The circuit must be installed inside the receiver and connected before the balance and volume controls of the receiver. The sensitivity can be adjusted by P1.

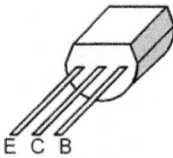

2SC3622
2SC3245
2SC3245A
2SC3248

*Transistor
equivalents*

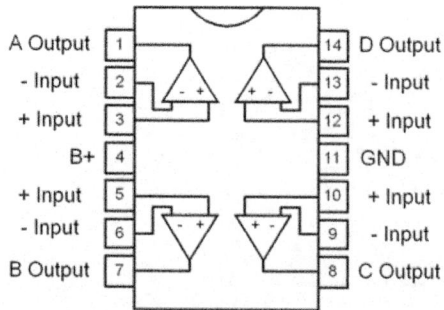

LM324
QUAD OPAMP

*LM324
equivalents:*

TDB0124
TDB0324
CA324
AN6564
TA75924

46 AUDIO FILTER

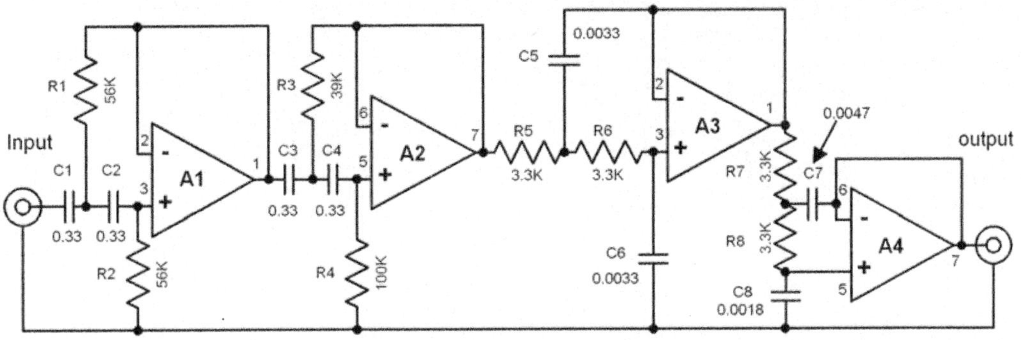

A1,A2 = IC1 = NE5532N (SE5532N)
A3,A4 = IC2 = NE5532N (SE5532N)

Diagram 46.0 Audio Filter

This is a bandpass filter for audio frequency band. It is also called a noise filter. It filters unwanted signals that are lower or higher than the audio frequencies. Its bandpass characteristic is achieved through the technique of cascading a low pass filter with a high pass filter. Both filters are second-order filters with a 24 dB/octave filter capability. The 3 dB cut-off frequencies are 11.8 Hz and 10.7 kHz.

The bandpass characteristic can be modified by changing the values of the capacitors and resistors. When you want to raise the bottom cut-off frequency, you must reduce the values of C1,C2,C3 and C4. On the other hand when the capacitance values are increased the bottom cut-off frequency goes down.

If you want to reduce the top cut-off frequency, you must increase the values of resistors R5 up to R8. If the resistance values are decreased on the other hand, the top cut-off frequency goes up.

Parts List

Resistors:
R1,R2 = 56K
R3 = 39K
R4 = 100K
R5,R6,R7,R8 = 3.3K

Capacitors:
C1,C2,C3,C4 = 0.33µF ceramic
C5,C6 = 0.0033 µF ceramic
C7 = 0.0047 µF ceramic
C8 = 0.0018 µF ceramic
ICs:
IC1,IC2 = NE5532N

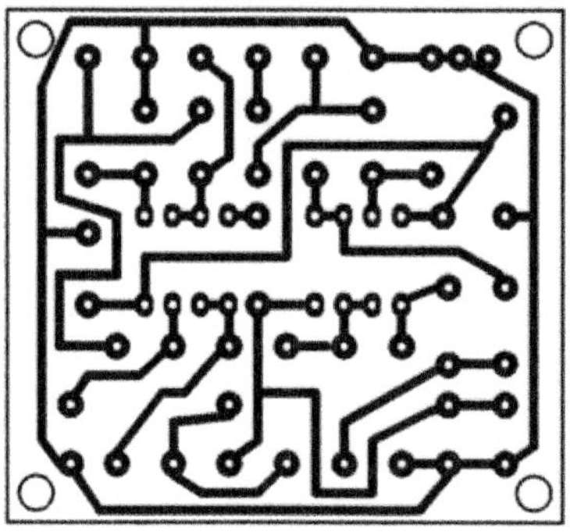

Figure 46.0 Printed Circuit Layout for the Audio Filter

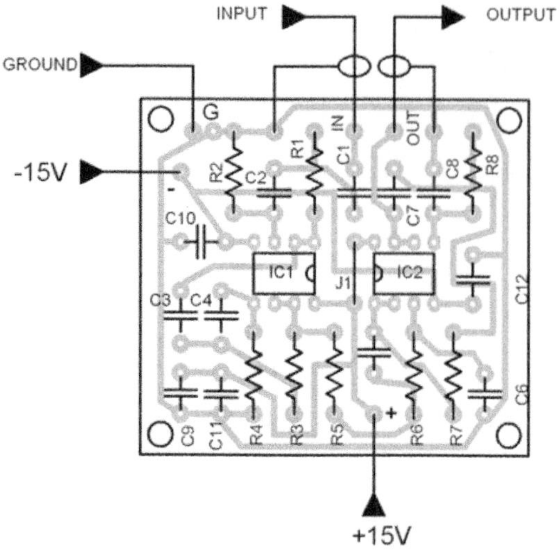

Figure 46.1 Parts Placement Layout & External Wiring

47 1-CHIP 40 WATT AMPLIFIER

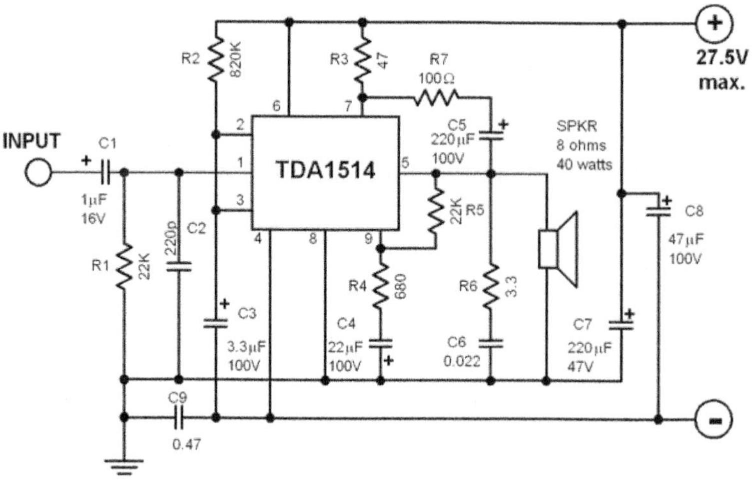

Diagram 47.0 1-Chip 40 Watt Amplifier

This is a compact and easy to build amplifier that uses one IC only but delivers 40 watts of audio power! It is ideal for amplifying audio from your mobile CD player or iPod. The chip being used here is the TDA1514. The best characteristic of this chip is high output power and robustness. It is available in a 9-pin SIL plastic package with a metal mount.

Its package has a heat resistance of less than 1.5K/W. This means that the heatsink must have a heat resistance of only 3.8K/W when the chip reaches its maximum power dissipation of 19 watts (at Ub = +/- 27.5 V, Tu = 50°C).

One can see from the diagram that only a handful of passive elements are needed to build the chip into a powerful audio amplifier. The power supply as supplied must be able to deliver a current of at least 3 amperes. The standby current consumption is about 60mA. The supply voltage must *never* go beyond 27.5 volts!

In building the circuit, keep the wires to the power supply and outputs as short as possible. The resistors R4 and R5 set the voltage gain at the feedback which, in this case, is between 20 and 46 dB. For a single channel amplifier (mono), a 80VA transformer (T1) should be sufficient. If you construct two channels (or stereo) amps, 120VA is recommended. Capacitors Cx and Cy should be at least 4,700mμF rated at 35V. It can be up to 10,000μF. Capacitors twice as large discharge slower, giving better peak power potential resulting to better power output. Feel free to increase the capacitance but take note that you may not get much additional benefit for the price involved. Make sure they are connected the right way around too, otherwise they will blow and might cause injury.

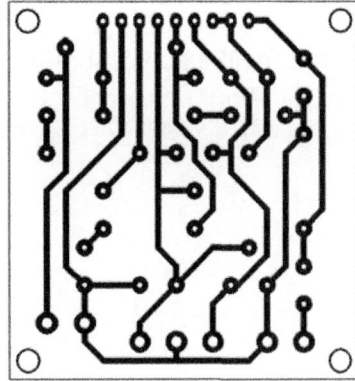

Figure 47.0 Printed Circuit Layout

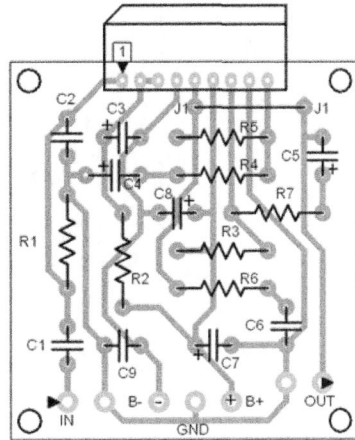

Figure 47.1 Parts Placement Layout

A 0.25mA fuse (F1) must be installed. If using a toroidal transformer, the fuse must be a time delay (slow blow) variant. Be sure to correctly earth the supply and any metal casing around it. The components on the earth and ground connections (Dx,Dy,Rx,Cz) form a loop breaker. This is recommended because it can eliminate those troublesome earth loops. Rx is a 5W or better wirewound resistor. The Cz capacitor must be rated for 250V AC. Do NOT use a 250V DC cap for Cz as it would fail if there ever was a fault causing mains current to flow to earth. Dx and Dy are rated at 250V/1A. If your local rules and regulations prohibit constructing this, omit all these components and connect the earth to ground but *NEVER* disconnect the earth lead... it could save your life or somebody else!

Power Supply:

Constructing the power supply for this amplifier is simple. As shown on diagram 47.1 you need to wire up a 18-0-18 (center tapped) transformer in order to get the recommended +/-25V. Be very careful since this construction involves mains wiring:

MAINS VOLTAGE IS VERY DANGEROUS.
DO NOT WIRE IT UNLESS YOU ARE QUALIFIED.
DEATH OR SERIOUS INJURY MAY RESULT.

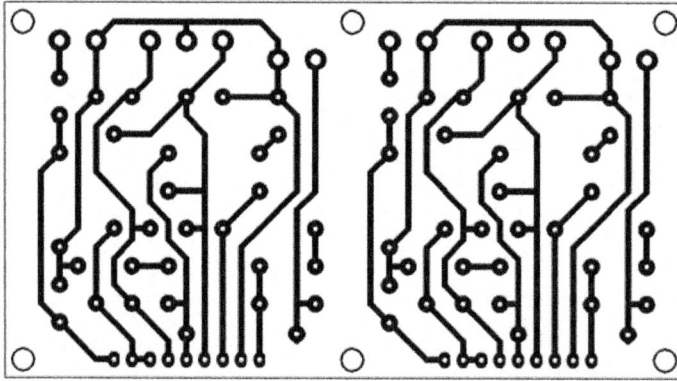

Figure 47.2 Printed Circuit Layout (stereo version)

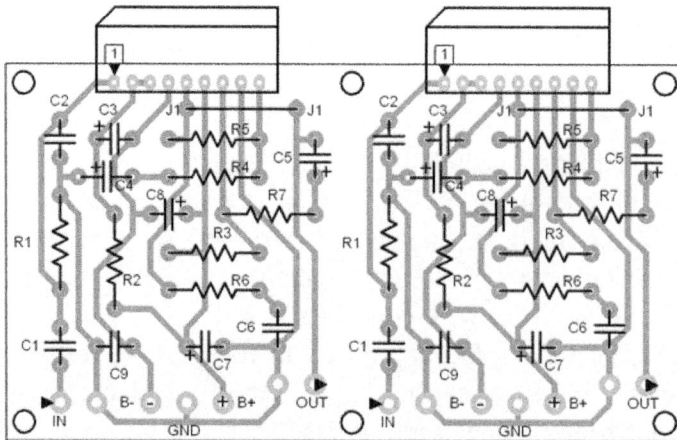

Figure 47.3 Parts Placement Layout (stereo version)

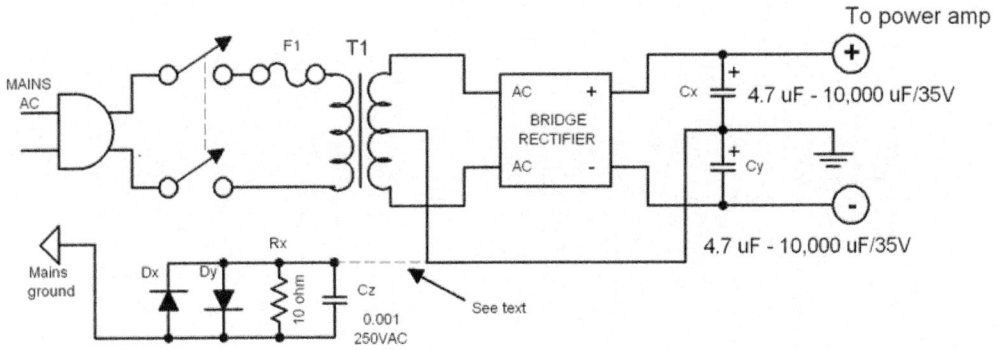

Diagram 47.1 Power supply for 1-chip amplifier

48 BALANCE INDICATOR

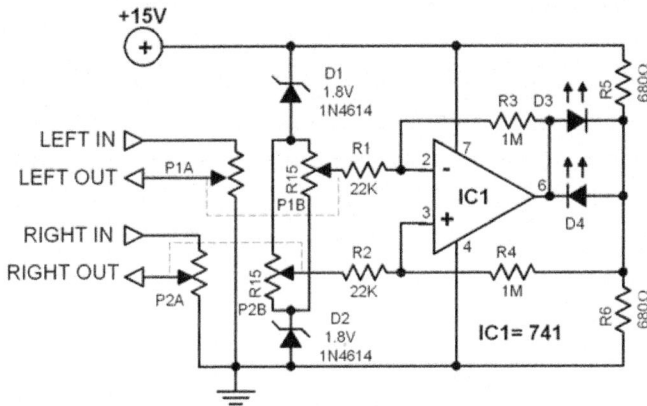

Diagram 48.0 Balance Indicator

Are you tired and frustrated of eternally trying to balance your stereo amplifier? If so, why not let electronics listen to the amplifier's output for you, and tell you whether the two channels are balanced or not?

Such automatic listener and balance analyzer is featured here. This simple electronic circuit provides an optical way of telling whether the two (left and right) channels are balanced or not.

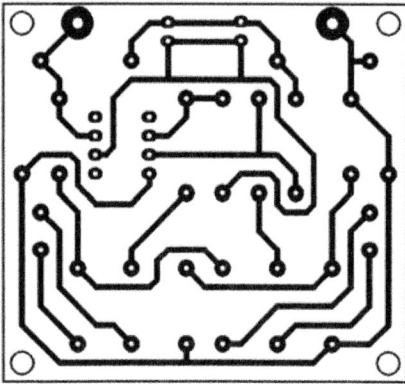

Figure 48.0   Printed Circuit Layout

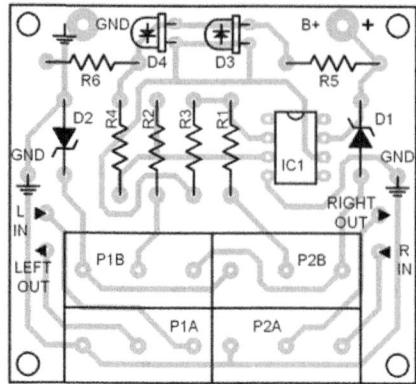

Figure 48.1   Parts Placement Layout

   The potentiometer pair P2a and P2b is mechanically coupled to each other as well as the pair P1a and P1b as shown by the dashed lines in the circuit diagram. There are also paired potentiometers available in electronics supply stores. Buying the ready made pots is highly recommended. They are accurate and will help you minimize your efforts in constructing this electronic circuit.

   After building the circuit, wire it properly. Important: turn down the volume of the amplifier before connecting this analyzer circuit. Afterwards, connect the input lines of the balance analyzer to the two output channels of your amplifier. Turn on the amplifier and feed a signal to its input. Ideally, the input signal must be a clean sine wave with a constant amplitude such as the one produced by a signal generator. Slowly, adjust the volume of one channel until a LED lights up. Adjust the volume of the other channel until the LED turns off.

   When the two channels are balanced, the LEDs remain turned off. Otherwise, one of the LEDs will light showing which channel is stronger.

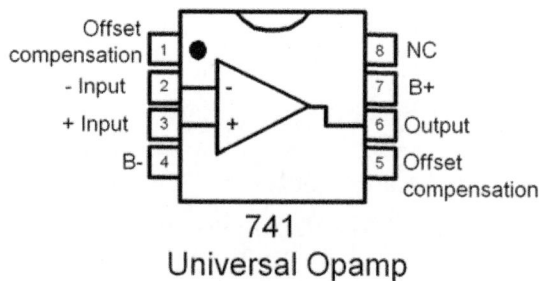

741
Universal Opamp

49   VIDEO AMPLIFIER

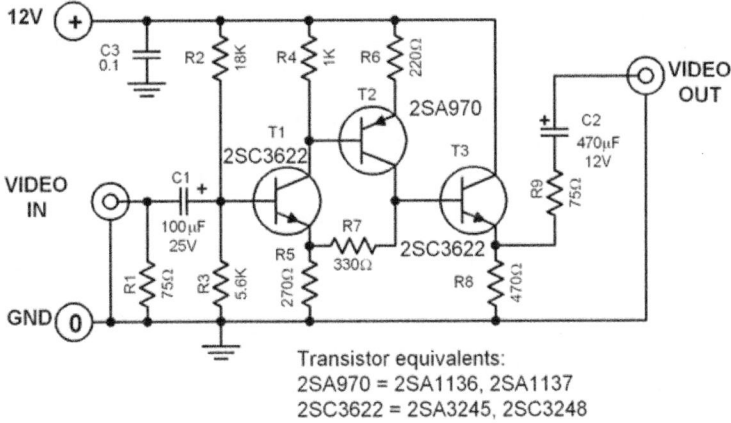

Diagram 49.0 Video Amplifier

This is a universal video amplifier that is very simple to construct. It does not need special components. By connecting this circuit in a long coaxial cable it compensates for the signal losses within the cable. The circuit is composed of 2-stage amplifier with an emitter follower as impedance converter.

The power supply must be stabilized properly to avoid signal interferences.

Technical Data

Amplification factor= approx. 2
Bandwidth=   over 20 MHz
Supply voltage=   12 V
Current consumption= 20 mA

Transistor equivalents

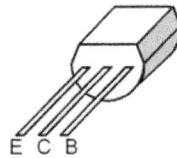

E C B

2SA970	2SC3622
2SA1136	2SC3245
2SA1137	2SC3248

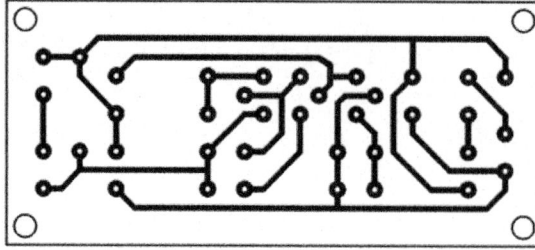

Figure 49.0 Printed Circuit Layout for the Video Amplifier

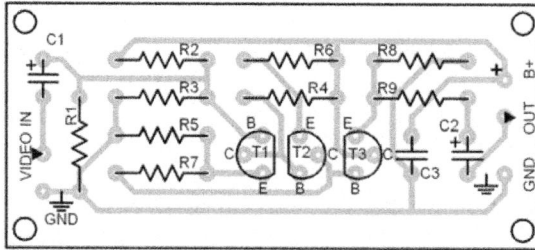

Figure 49.1 Parts Placement Layout for the Video Amplifier

50 E-GUITAR PREAMP

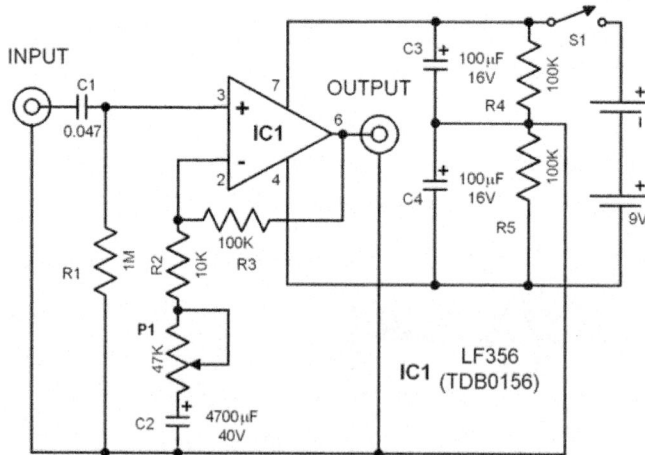

Diagram 50.0 E-Guitar Preamp

Simple but powerful. This preamp circuit is commonly used to raise the output level of an electric guitar to enable it to drive vacuum tube amplifiers. Overdriving the vacuum tube amplifier is sometimes necessary to achieve a certain "guitar sound". Since the normal output level of a guitar is not enough for this purpose, the small circuit featured here is just the right thing to do the „synthesizer" trick.

Technical Specifications

Amplification factor:
-set by   R2+R3+P1/(R2+P1)
Input impedance: 1 MΩ
Supply voltage: 9 volt
Current consumption: 5 mA

This preamp amplifies the guitar output to such a level that the input stages of the final amplifier will be over driven to their limiting level. The amplification factor of the preamp is adjustable from 3 to about 11.

As you can see clearly, the circuit is very simple. A single LF356 IC gives the signal the needed boost. The input impedance (1 MΩ) is practically determined by R1 since the IC has FET inputs. One megaohm is an ideal value since it is the impedance of most guitar pickups. The voltage divider network R4,R5,C3 and C4 provides the IC with a symmetrical +/-4.5 volts. Due to its low current consumption, the circuit can be powered with small batteries and housed in a very compact plastic case.

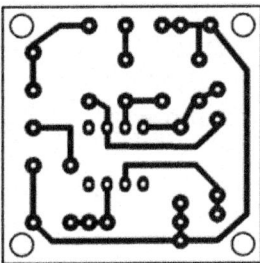

Figure 50.0
Printed Circuit Layout
for the E-Guitar Preamp

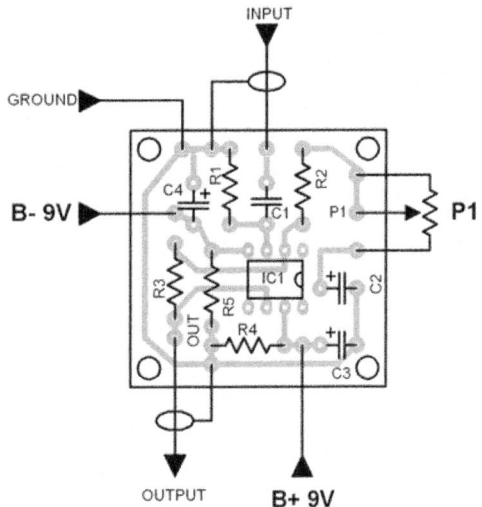

Figure 50.1
Parts Placement Layout
& External Wiring

51   HIFI STEREO PREAMP

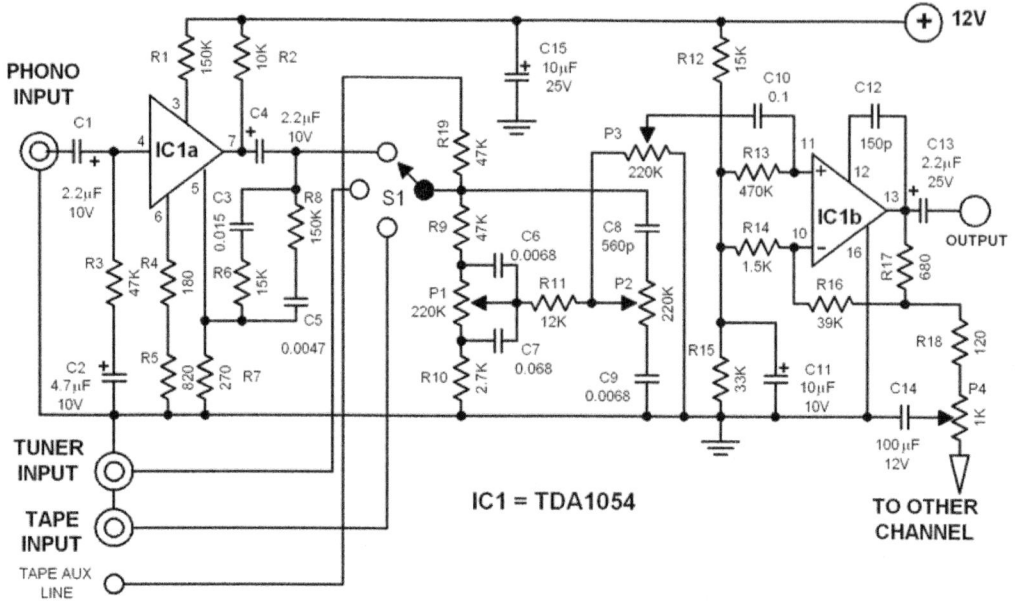

Diagram 51.0   HIFI Stereo Preamplifier

Nowadays, there is an abundance of ready to use final power amplifiers in electronic stores. Many of these devices have great electrical data in terms of distortion factor, power output and frequency bandwidths. However, no matter how good these data are, these power amplifiers are still of no good use without a good preamp stage.

Lately, a new breed of preamp IC's came to the market. These IC's allow the hobbyist to build low noise preamp circuits with little or no problem. One of these IC's is the TDA1054 from SGS. It is a 16-pin DIL package and integrates two separate preamp circuits. The first half of the circuit (IC1a) is a preamp with an input sensitivity of 3 millivolts. It has a frequency correction composed of R6,R8, C3 and C5. The bass signal coming from the phono input is amplified while the high signal is attenuated.

Switch S1 allows the selection of the input signal source. The potentiometers P1 and P2 are part of a double potentiometer. They control the bass and high tones. There is no risk of overdrive in this circuit due to the passive nature of the sound control.

Potentiometer P3 controls the volume of the signal fed to the second part of the circuit (IC1b) which functions as an operation amplifier. P4 controls the balance between the left and right channels (one channel not shown on the diagram). At the middle setting of P4, both channels have a gain of 24. If P4 is set to one extreme end, the gain difference between the two channels is about 12 dB.

Diagram 51.1 shows the voltage regulator circuit for the preamp. It uses the 7812 IC and supplies 12 volts.

Input sensitivity by a frequency of 1 kHz and output of 775 mVeff:

Phono = 3mV/50k
Tuner = 220mV/50k
Tape   = 220mV/50k

Maximum output voltage = 2.5 eff.
Balance range   = 12 dB

Sound adjustment:
low = +/- 13 dB (100 Hz)
high = +/- 13 dB (100 Hz)

Distortion = <0.05% (f = 1 kHz, output voltage = 775 mVeff)

Bandwidth = 20 Hz ... 24 kHz

Signal to noise ratio = >65 dB (output voltage = 775mVeff)

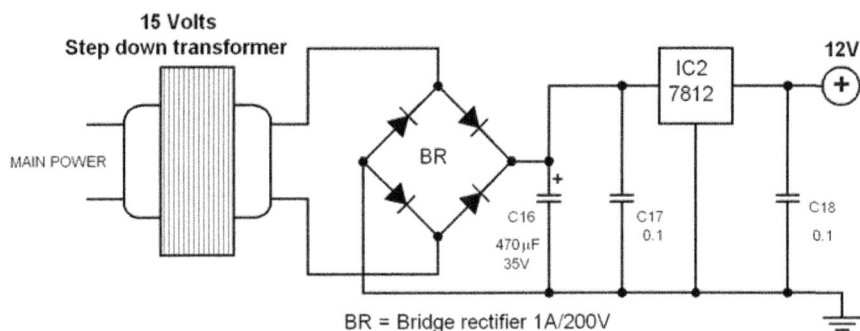

Diagram 51.1 Regulated power supply for
the HIFI stereo preamplifier

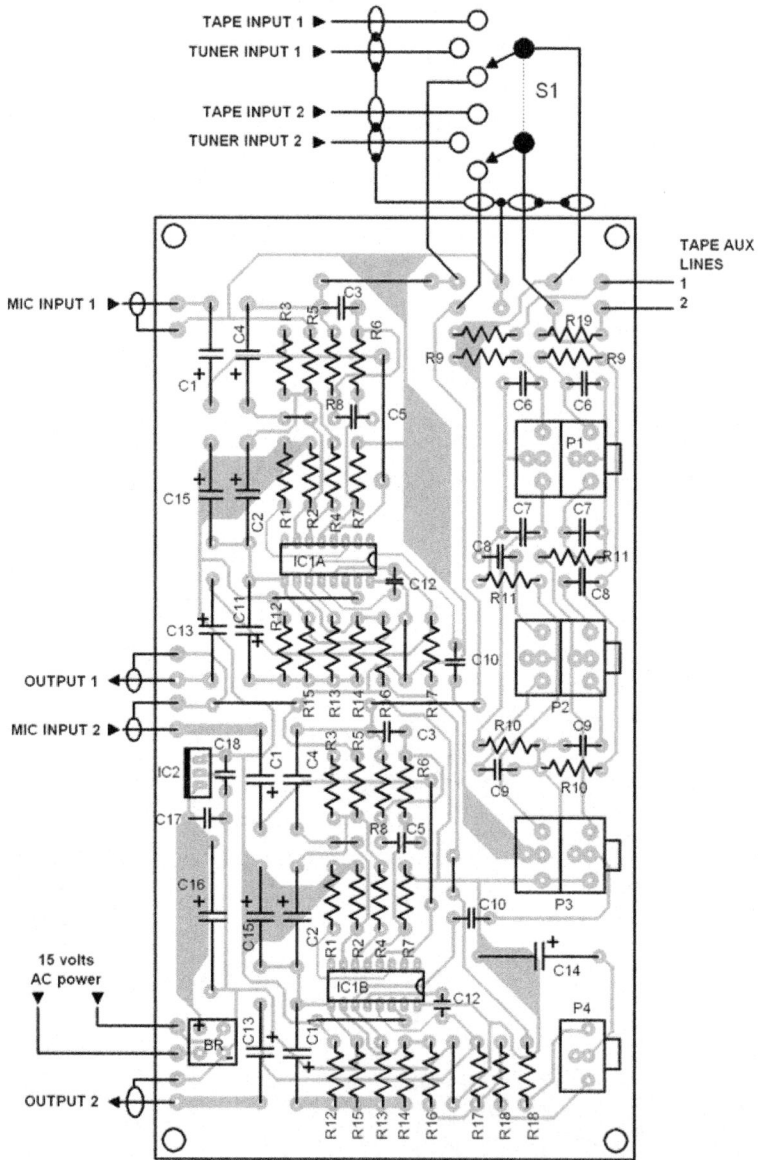

Figure 51.0 Parts Placement Layout for the HIFI Stereo
Preamplifier with external wiring guide

Caution! Danger of electrocution!
Extreme shock hazard! You are working with
a line voltage of 220 Volts AC.

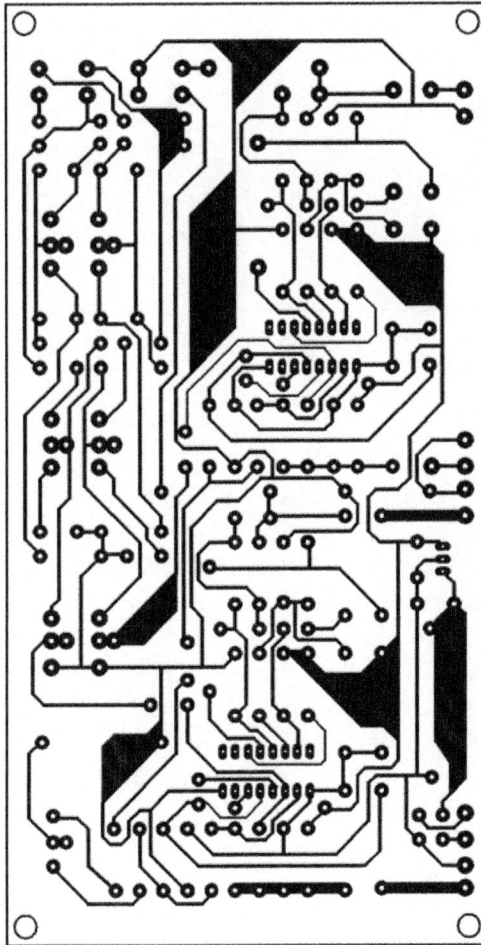

Figure 51.1 Printed Circuit Layout for the HIFI Stereo Preamplifier

52 HIFI AUDIO MIXER

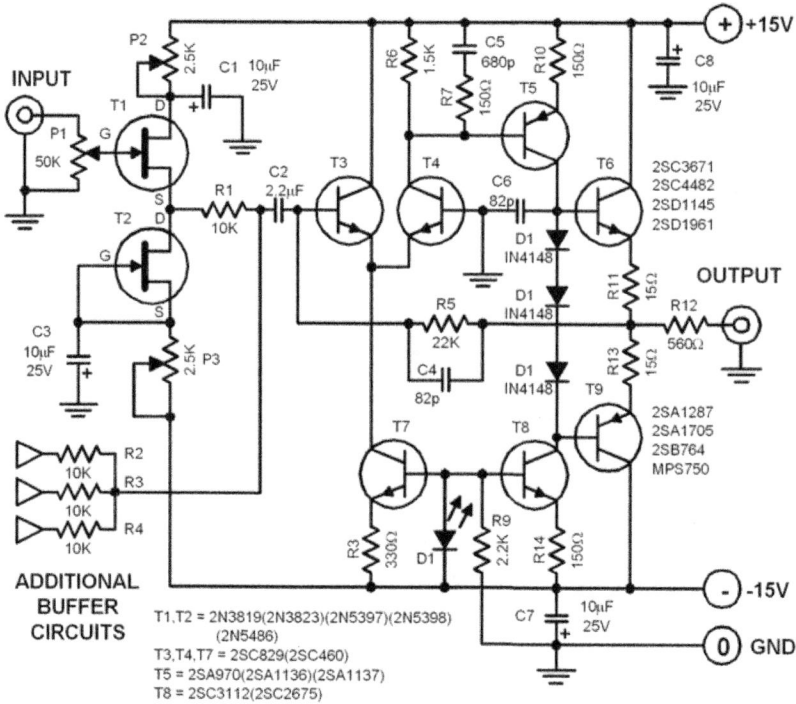

Diagram 52.0   HIFI Audio Mixer

   High fidelity (HIFI) mixers deliver high dynamic and low noise characteristics but they are expensive. Conventional mixers on the other hand are constructed with cheap op-amps that are somewhat noisy. The noise problem can be avoided by buffering the inputs and instead of using op-amps, discrete components are used for the amplifier stages. These design considerations are applied in the circuit featured here. Transistors T1 and T2 are the buffers. The input impedance is dependent on the setting of P1. Transistors T3 up to T8 build the amplifier stage and are HF type transistors. HF transistors have lower noise factor than their AF counterparts.

   The described circuit is designed not only for mono inputs. It can also be expanded to process stereo inputs. To do this, just construct two identical circuits and use them as separate modules for both left and right inputs. One thing to be very careful though in constructing the stereo modules, make sure that the wirings are not mistakenly crossed between the two identical circuits.

Technical characteristics: Frequency range = (-3 dB) 10 Hz ... 80 kHz
Signal to noise ratio (9V output/20 kHz bandwidth) = 100 dB with 10 buffer stages
Maximum output signal = 12 Vpp

   The buffer circuit must be duplicated for every additional input channel and connected to C1 in the amplifier stage.

53 SUBSONIC FILTER

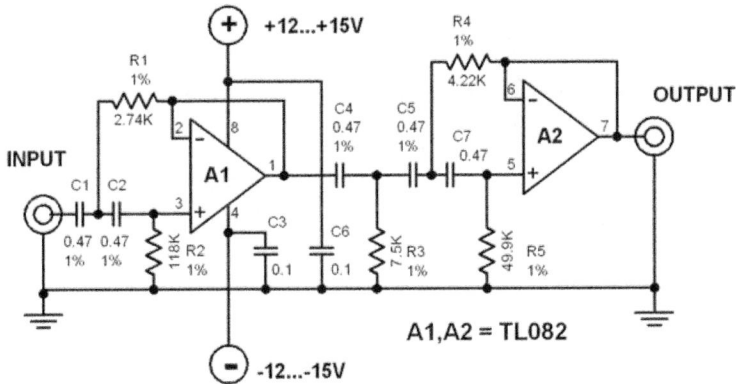

Diagram 53.0 Subsonic Filter

A subsonic filter is mostly used in low frequency applications. Most often, the very low frequencies cause distortions in music recording or reproduction. This happens independently from the applied recording medium (e.g. CD, tape, vinyl record, etc). To remove this unwanted low frequencies, subsonic filters like the electronic circuit featured here is used.

The circuit is actually an active Tschebyscheff high pass filter of the fifth order. It has a cutoff frequency of 18 Hz. This filter has a very sharp cutoff character. Frequencies below 10 Hz are attenuated by more than 35 dB. The typical Tschebyscheff resonance of 0.1 dB does not cause any distortion. The phase shift is negligible.

If the circuit is applied to a stereo unit, build an exact duplicate for the other channel. Any difference could cause noticeable phase shift in the output signal. It is very important that the capacitors C1, C2, C4, C5, and C7 are of exactly the same values in both channels.

The current consumption is around 5 mA. The maximum useful frequency is 3 MHz. In connecting the filter, remove the decoupling capacitor from the front end module to avoid distorting the signal.

54 SELECTIVE FILTER

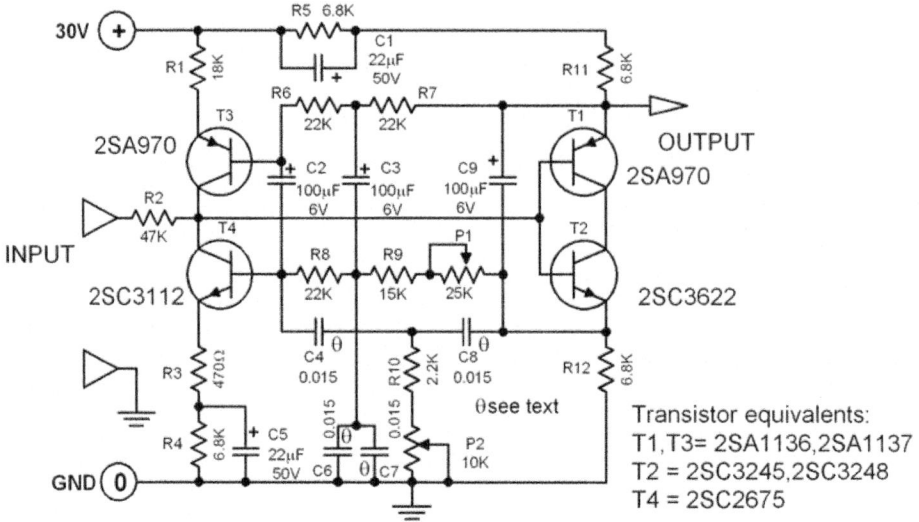

Diagram 54.0 Selective Filter

This circuit is an audio filter with a so called double T network. It is a narrow bandpass filter. It can be set at a certain frequency (fo) with trimmer potentiometers. All frequencies other than the fo are attenuated. The circuit has a complementary emitter follower composed of T1 and T2 which are controlled through R1. The emitter of T1 or T2 is the output. The output is connected to a balance amplifier T3/T4 through the double T network of P1,P2,R6 to R10, and C4,C6,C7,C8 (marked in the diagram). The amplification (A) of the balance amplifier is:

$$A = \frac{2R2}{R1} \equiv \frac{2R2}{R3}$$

$$fo = \frac{1}{RC}$$

$$R6, R7, R8 = 2R$$
$$R9 + P1 = 2R$$
$$R10 + P2 = R/2$$

Potentiometers P1 and P2 set the maximum output at the frequency fo.
When $R = 11K$ and C4,C6,C7,C8= 0.015 , the fo is around 1000 Hz (1kHz).

55   INTERCOM

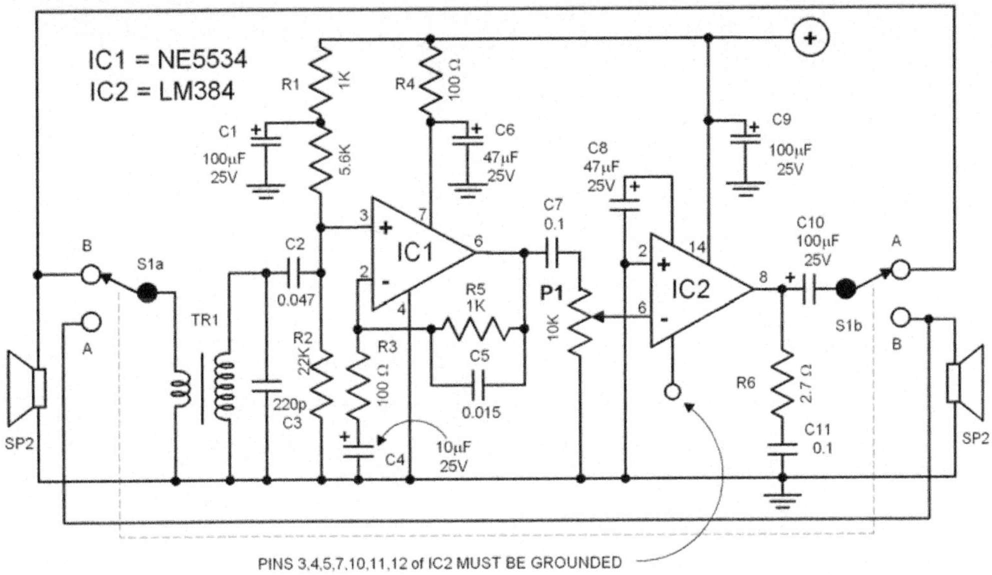

IC1 = NE5534
IC2 = LM384

PINS 3,4,5,7,10,11,12 of IC2 MUST BE GROUNDED

Diagram 55.0 Intercom

If you compare this intercom with a modern FM and wireless designs, you will certainly say it is obsolete. Well..not quite right. The circuit fulfills its task - to provide a reliable communication line - and is very simple to construct. And that's what really counts. The circuit is made up of an amplifier, two switches, and two loudspeakers. If more stations (speakers) are needed, additional switches are just incorporated into the circuit. IC LM382 is used as the final amplifier. It delivers nearly 2 watts of audio power by 15 volts supply voltage. Amplification is preset to 50. C9 serves as supply bypass. Another IC is used to boost the input signal before it is passed into the LM384. IC2 is an op-amp with an amplification factor of 11. The frequency bandwidth is 160 Hz to 10 kHz. These values were chosen since the device is to be used only as intercom and not as HIFI amplifier.

To achieve the best performance, use speakers that can also function as microphones. The intercom must be constructed inside a box so that the microphone function is optimum. Since the impedance of most speakers are quite low, an impedance converter is needed to maintain a good audio quality. This is done by the transformer TR1. TR1 is a normal 220/6V step-down transformer. The 6V winding of the transformer is connected to the speaker, and raises its impedance to about 10 kilo-ohms. To reduce signal loss in the transformer, a 4.5 Watt transformer is used.

In constructing the intercom, house the power supply unit separately from the main circuit to avoid interference. C1 suppresses HF interference. The current consumption is about 210 mA with 1.8 watt output.

56 AUTOMATIC VOLUME CONTROL

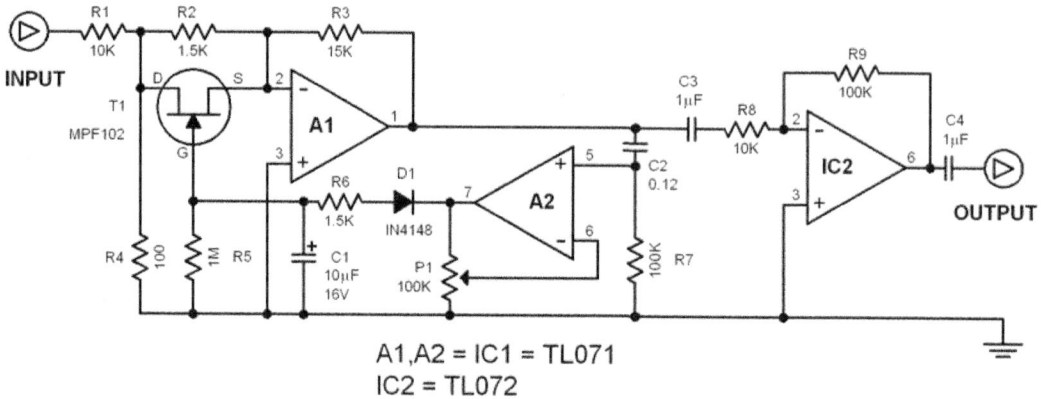

Diagram 56.0 Automatic Volume Control

The function of this electronic circuit is to amplify weak signals without distorting its dynamic compression. The amplitude differences in the signal are levelled off and the disturbing effect vanish. With this technique, overcompensation in the volume is avoided. The circuit is used to automatically control the volume levels of cassette recorders, audio tape recorders, amplifiers, or radio devices.

The circuit works this way: The FET (T1) is used as a variable resistor. The resistance between the drain and source of T1 can be between 150 ohms and infinite. It is parallel to R2 and together with R3 controls the gain of op-amp A1 (around 20 dB). The following op-amp A2 is wired as an amplifier with P1 as its gain control. The negative part of the output signal coming from A2 is rectified and fed to the gate of T1. Small variations in the signal amplitude does not influence the amplification since the FET has a short delay caused by R6. The opposite effect is also slow because of the discharge time of the C1. Both effects result to a smooth regulation of the signal amplitude.

The signal voltage at the gate of T1 must be as low as possible to influence the drain-source resistance. To achieve this, the voltage divider R1,R4 is set at the input line to attenuate the signal by 40 dB. This technique enables signals up to 1 Veff to be processed without a problem and with distortion levels below 0.5%.

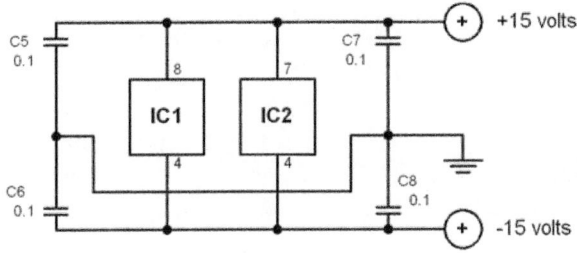

Diagram 56.1 Decoupling capacitors at the ICs' power supply lines.

The signal-to-noise ratio is over 70dB by an input voltage of 1 Veff. The losses at the attenuator is compensated by the amplification through A1 and IC2. The high pass filter made of C2,R7 prevents the bass signals from influencing the regulation. With proper dimensioning of these components, the cutoff frequency of this filter can be adapted to individual needs. Signals below the threshold set by P1 are amplified by around 18 dB.

The circuit needs a symmetrical supply of +/- 15 volts and consumes around 7 mA.

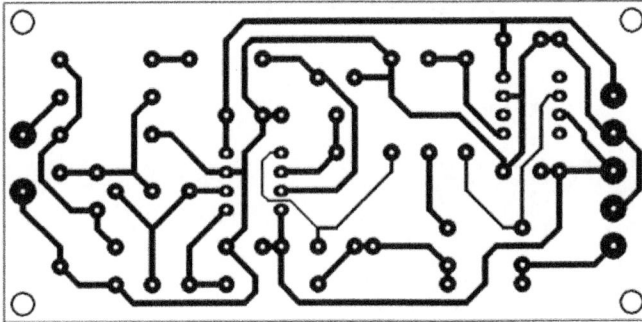

Figure 56.0 Printed Circuit Layout

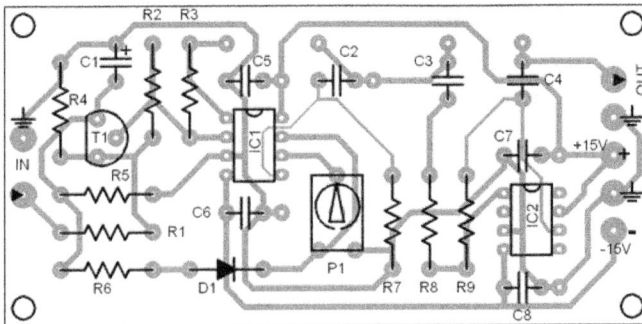

Figure 56.1 Parts Placement Layout

57   MUSIC CHIP

Diagram 57.0   Music Chip

This 1-chip module can generate nine different melodies. It is commonly used as a doorbell, an acoustic signalling device or as a telephone "wait" melody.

Once switch S1 is pressed, the IC plays the melodies one after the other. It will stop after the last melody when the JP1 terminal is connected to ground by a jumper. Alternatively, when JP1 is connected to the +V, the IC will play continuously as long as the switch S1 remains closed. Combining several pins produces different effects as shown in Table 57.0. A single battery can be used to power this module since its current consumption is negligible.

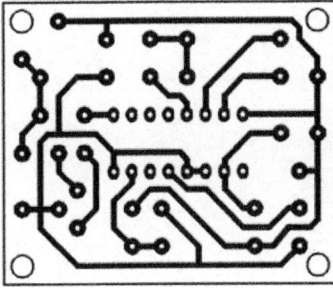

Figure 57.0 Printed Circuit Layout

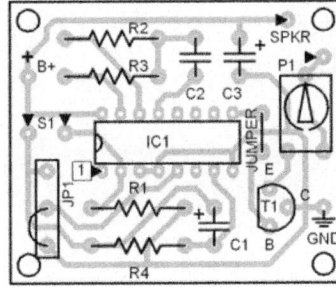

Figure 57.1 Parts Placement Layout

Table 57.0				
Pin 2	Pin 3	Pin 4	Pin 5	Effect
0	x	x	x	Standby
1	0	0	0	first-last-stop
+ Pulse	0	0	1	first-last-repeat first
+ Pulse	1	0	0	current-stop
1	1	0	1	repeat current
1	0	+ Pulse	0	next-last-stop
1	0	+ Pulse	1	next-last-repeat first
1	1	+ Pulse	0	next-stop
1	1	+ Pulse	1	next-repeat same

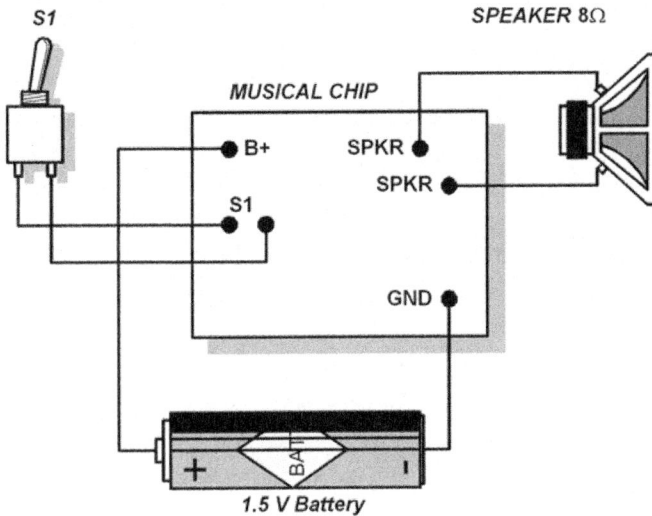

Figure 57.2 External Wiring Layout

58 AUDIO TESTER

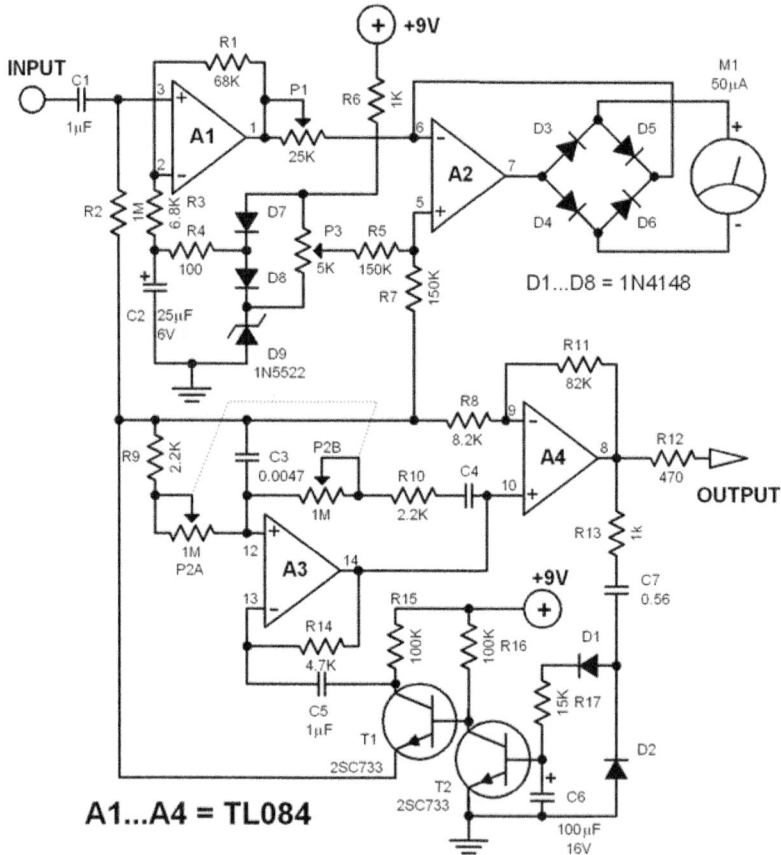

INPUT

+9V

R1
68K

P1

R6
1K

A1

P1
25K

R2
1M

R3
6.8K

R4
100

D7

D8

C2 25μF
6V

D9
1N5522

A2

D3   D5

D4   D6

M1
50μA

D1...D8 = 1N4148

P3   R5
5K   150K

R7
150K

R11
82K

R8
8.2K

A4

R12
470

OUTPUT

R9
2.2K

C3
0.0047

P2B

R10
1M

C4
2.2K

A3

1M
P2A

R14
4.7K

C5
1μF

T1
2SC733

T2
2SC733

R15
100K

R16
100K

+9V

R13
1k

C7
0.56

D1

15K R17

D2

C6
100μF
16V

A1...A4 = TL084

Diagram 58.0 Audio Tester

This circuit is a millivoltmeter plus a signal injector in one package. A simple milli-voltmeter combined with a sine wave generator is an ideal device to test audio modules. As shown in the diagram, the op-amps A1 and A2 make the millivoltmeter. The lower op-amps A3 and A4 make the sine wave generator. The frequency bandwidth is from 150 Hz to 20 kHz.

Since this electronic circuit is powered by a 9 volt battery, the supply line voltage for op-amp operation must be divided by two. This is accomplished by the zener diode D9. Resistor R6 is the load resistance for D9. The reference voltage is taken from the D8/D7 junction through the R4/C2 line. This reference voltage is about 5.3 volts. The constant voltage through the two diodes is tapped by P3 and used as variable offset bias for A2. This offset is used to calibrate the millivoltmeter to zero.

Let us check the millivoltmeter. The input signal enters the plus pin of A1 via the high pass filter created by C1/R2. The input impedance at this point is 1 mega-ohm. The maximum allowable input level is 50mVeff. To raise this level, one can either add a voltage divider circuit at the input or reduce the gain of A1 by substituting R1 with a lower value (e.g. for R1 = 6.8 Kohm, the gain is 2 and the input sensitivity is 275mVeff).

The full deflection of the meter M1 is set with P1. The op-amp A2 combines with the diodes D3,D4,D5,D6 to create a full wave rectifier.

The signal generator is basically a wienbridge oscillator with P2 and the capacitors C3 and C4. To stabilize the sine wave, the output signal is tapped from the buffer amplifier A4, rectified via D1,D2,C6,C7. Eventually it is fed to the minus input of A3 via the buffer transistors T1 and T2. With this technique, a relatively stable amplitude of about 2Vpp is generated.

The meter M1 can be any type from 50 µA up to 1 mA. However, the value of P1 in the diagram is chosen for a 50µA meter. To use other current values, convert the value of P1 linearly. For example, for a 500µA meter use 2.5Kohm for P1.

Calibration: take the reference voltage and reduce it to about 45mV by using voltage dividers. Feed this lower voltage level to the plus input of A1 and turn P1 until the meter displays „45" mV.

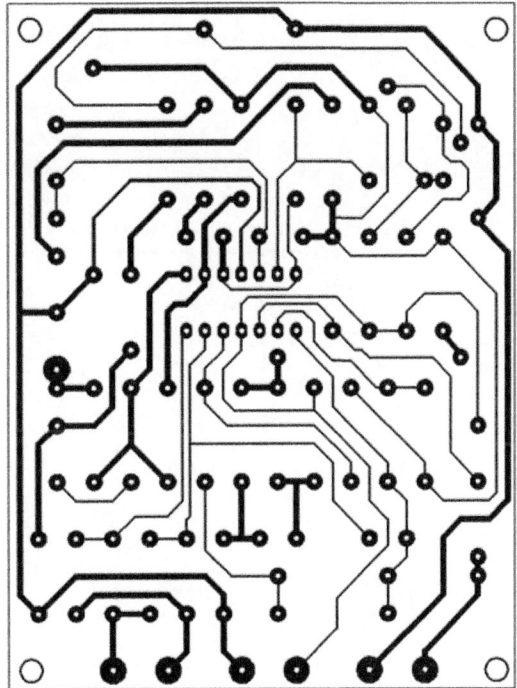

Figure 58.0 Printed Circuit Layout

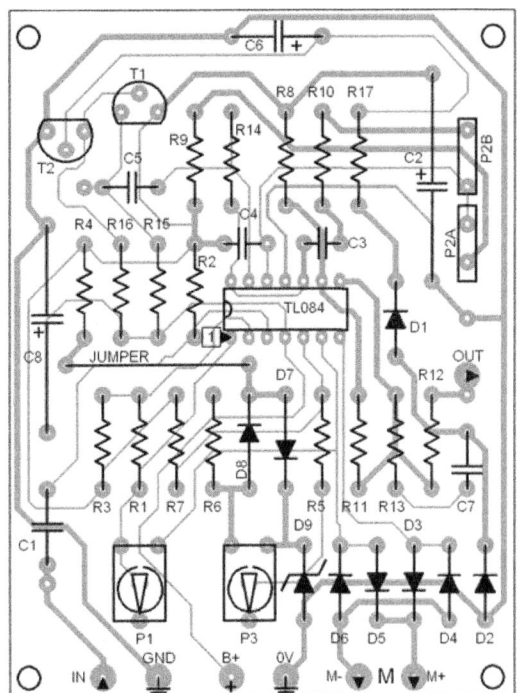

Figure 58.1 Parts Placement Layout

59   ULTRASOUND RECEIVER

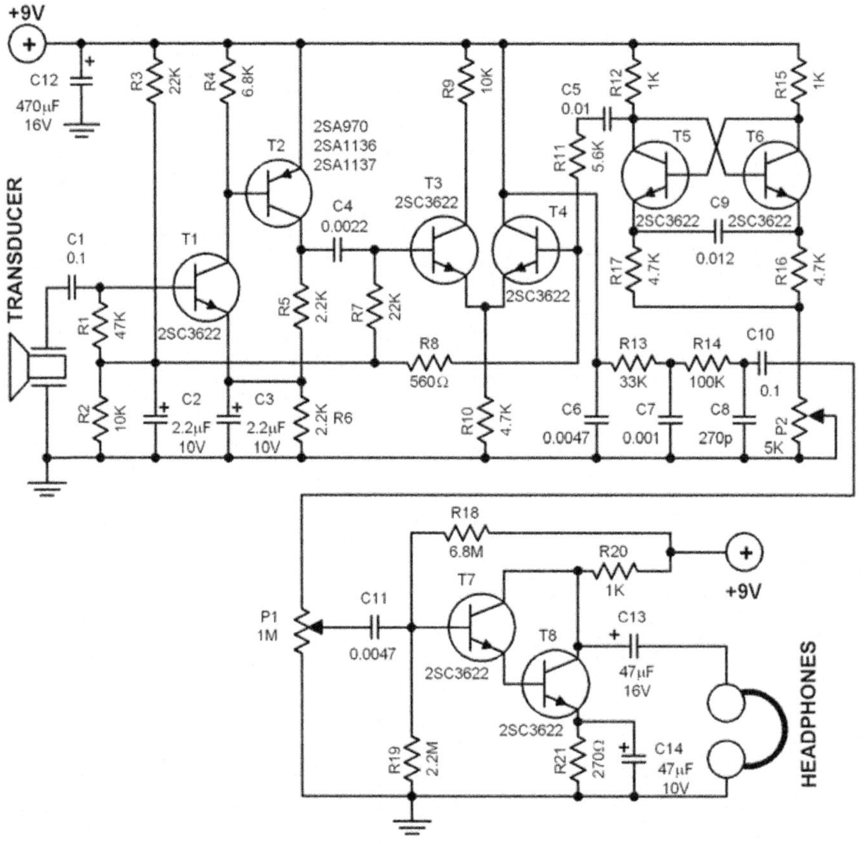

Diagram 59.0   Ultrasound Receiver

The audio frequencies between 15 kHz and 18 kHz is normally not audible to humans. This audio range is also called ultrasound. In order to hear it, it must first be converted to a frequency range which is audible to humans.

The circuit featured here mixes the ultrasound with a frequency generated by a BFO circuit. The result is a mixture of several frequencies from which one (the difference) can be clearly heard. The undesired signals are filtered out by a 4 kHz low pass filter. The transducer is a special ultrasound sensor. Portable batteries can be used to power this circuit.

60 FM STEREO NOISE SUPPRESSOR

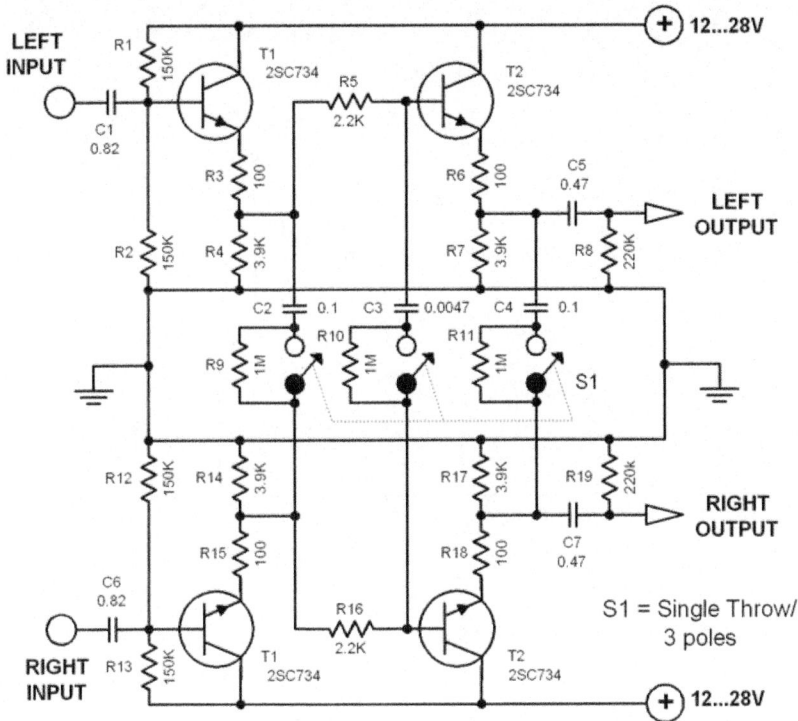

Diagram 60.0 FM Stereo Noise Suppressor

Sometimes reception of far FM stations is bad but improves when you switch the radio receiver to mono. The cause of this effect is that in stereo reception the biggest part of the noise in the right channel is in opposite phase with the noise in the left channel. Once the two channels are switched together (mono), these noise signals cancel each other. The result is a clean and relatively noise free mono signal. To use this effect in improving the signal but keeping the stereo character, the two channel should be switched together only in higher audio frequencies. This is done with the help of the above featured circuit.

The circuit is built in two identical channels each composed of two emitter-follower circuits. The switch S1 can connect the two channels together at three points in the circuit. The circuit functions as a bridge between the left and right channel for opposing signal components above 8 kHz. These signals get „short-circuited" cancelling the noise. Otherwise, all other frequencies get through to the output via their respective channels. If you want to lower the „short-circuit" frequency to 4 kHz, just double the values of C2, C3, and C4.

61   INTERCOM

T1,T3 = 2SA970(2SA1136)(2SA2237)
T2,T4 = 2SC3622(2SC3245)(2SC3248)

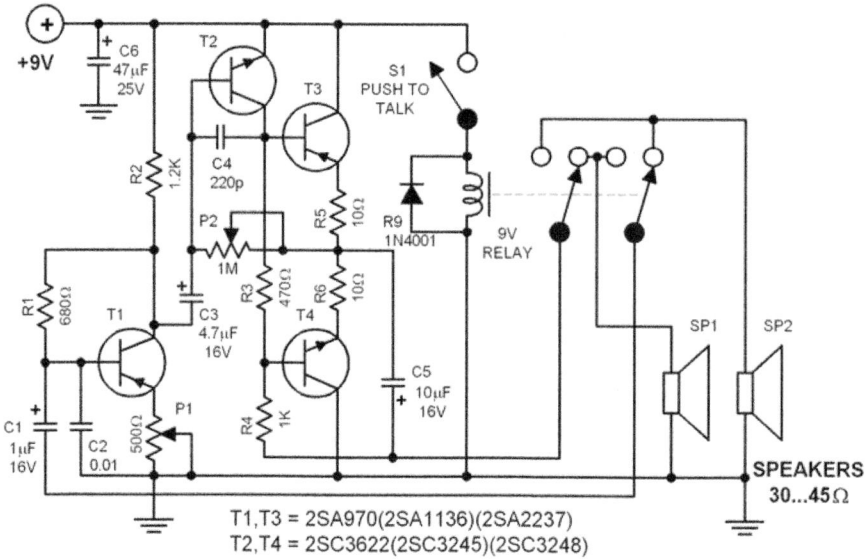

Diagram 61.0   Intercom

This intercom uses an ordinary speaker to function as its microphone. Once the button is pressed, the relay toggles, and the speakers interchange their functions. One works as a normal speaker and the other one works as microphone.

The circuit is composed of a complementary final amplifier (T3 & T4) and two voltage preamplifier (T1 & T2). Volume is controlled through P1.

The voltage at the common junction of resistors R5, R6 & P2 must be set to one half of the supply voltage.

A lower impedance speaker with a series resistor can replace the suggested type, but will deliver lower sound quality.

Parts List:

R1 = 680Ω		C1 = 1µF/16V
R2 = 1.2K		C2 = 0.01/50V ceramic
R3 = 470Ω		C3 = 4.7µF/16V
R4 = 1K		C4 = 220p/50V ceramic
R5 = 10Ω		C5 = 10µF/16V
P1 = 500 Ω		C6 = 47µF/25V

T1,T3 = 2SA970(2SA1136)(2SA1137)
T2,T4 = 2SC3622(2SC3245)(2SC3248)
Relay (DPDT) 9V
Speakers (30...45W) = 2 pcs.
Switch (SPST) = 1 pc.

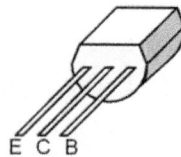

2SA970   2SC3622
2SA1136   2SC3245
2SA1137   2SC3248

E C B

Transistor equivalents

Page 114

62   SPEAKER BALANCE

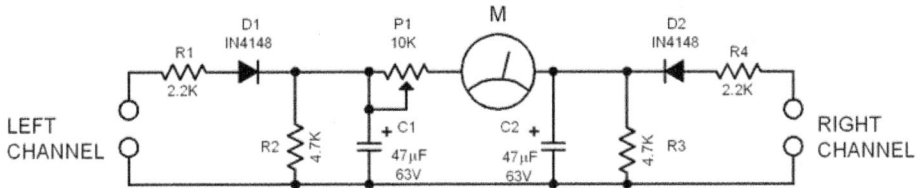

Figure 62.0   Speaker Balance

When using a stereo amplifier, there are many possible mechanical problems that could influence the amplifier's output performance. One would wish to avoid them but they happen more often than one thinks. Worse, the unwanted effects are usually very subtle that they don't even get noticed at once. The usual culprit is the mechanical potentiometers which are not rarely un-sync in their stereo resistance pads. This results to unbalanced volume levels on the audio channels. Most good stereo units have a balance control to compensate for this imbalance.

In order to know exactly the imbalance and be able to correct it, one needs the speaker balance indicator circuit featured here. In using the circuit, its left and right input channels are connected to the corresponding speaker outputs of the amplifier. Identical signals are then fed to the amplifier's stereo inputs. The best signal would be a pure sine wave from a signal generator. When a signal of exactly the same amplitude is coming out from both speaker outputs, the meter M of the speaker balance circuit will stay at its zero setting. The meter M is a type with its zero setting at the middle of the scale. This enables one to see at first glance when one channel is louder than the other channel, e.g. if the meter swings to the left, it means that the signal amplitude at the left channel is higher than at the right channel. To correct the imbalance, one has to turn the amplifier's balance control until the meter goes back to the middle of the scale.

Calibrating the speaker balance indicator circuit: Inject a signal in one channel and set the amplifier's volume to maximum. While doing this, turn the potentiometer P1 until the meter swings to end of the scale in the direction of the channel, e.g. left end of scale for left channel and right end of scale for right channel.

63 AUDIO DELAY LINE

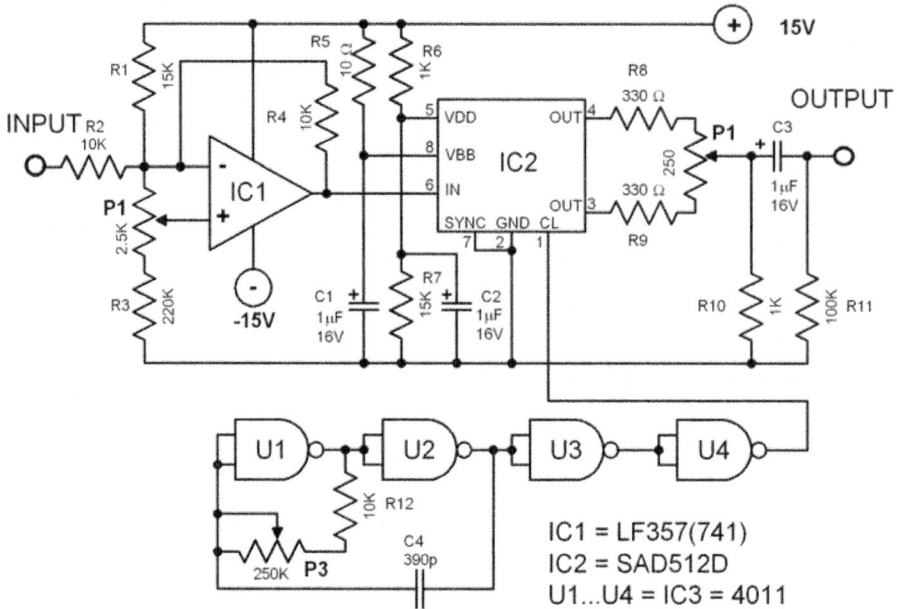

Diagram 63.0 Audio Delay Line

There are several different techniques in delaying an audio signal to achieve an echo effect: Hall and echo devices, studio-effect installations, and chamber simulations, etc. The most portable and newest method is the so-called bucket brigade delay (BBD). The delay circuit featured here uses SAD512. This IC has 512 bucket-brigade memory elements with an integrated clock. The analog input signal is converted to a DC voltage by IC1. The four NAND gates U1...U4 function as a clock generator and can be controlled by P3 from 10 kHz to 300 kHz.

The clock frequency of the bucket-brigade memory is controlled between 5 kHz to 50 kHz since the internal clock of the IC is divided into two. The delay of the circuit can be computed through this formula: The delay can be adjusted from 51.2 mS up to 5.1 mS. The maximum frequency of the audio signal that can be processed by the IC is equal to one half of the clock frequency - that is between 2.5 kHz and 25 kHz.

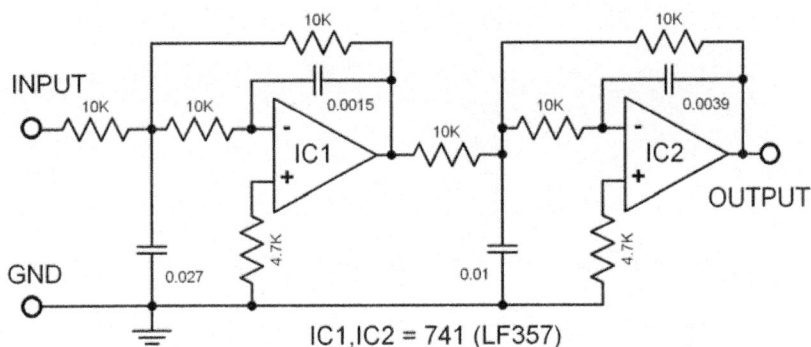

Diagram 63.1 Low-Pass Filter for the Audio Delay Line

The last output signal and the second to the last signal are mixed together by P2 so that the clock frequency can be suppressed as much as possible. Normally, the suppression of the clock signal by P2 is not enough. Therefore the low-pass filter shown in diagram 63.1 must be used to obtain best results. This filter circuit is a fourth order Butterworth filter. The 3 dB cut-off frequency is around 2.5 kHz.

When the frequency of the input signal is higher than one half of the clock's frequency, the circuit will produce unwanted mix-products. These are the so-called "fold-over" distortions. To avoid this kind of distortion, a second lowpass filter should be used in front of the delay circuit. This filter is identical to the one shown in diagram 63.1.

Formula for T-delay

$$T\ delay = \frac{n}{2fc} = \frac{512}{2fc}$$

Note that **n** means the number of "bucket" elements.

---

64 FULL DUPLEX AUDIO LINE

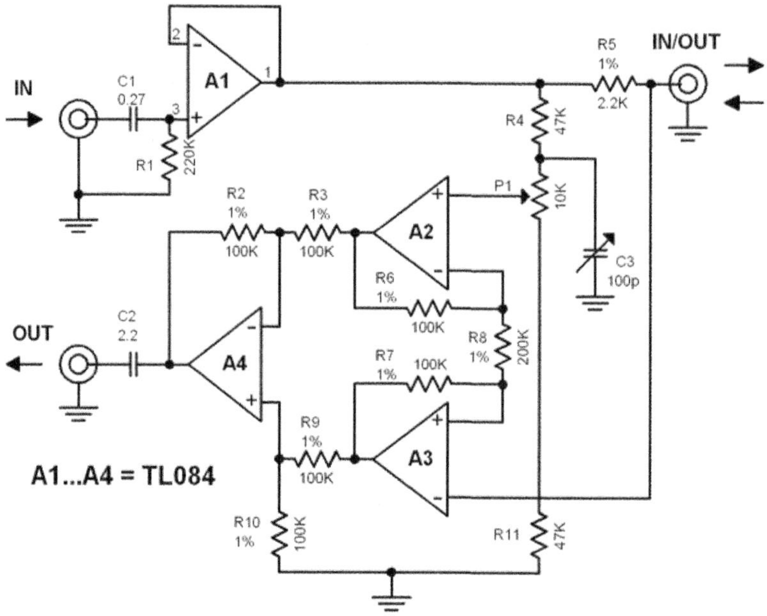

Diagram 64.0 Full Duplex Audio Line

This circuit enables two audio signals in opposing directions to flow simultaneously through a common twisted pair line. It does its job without complicated communications technology. This technique, called full duplex, is not really new. It is common in telephone technology. However, the telephone uses a carrier signal. This circuit on the other hand does the trick without a carrier signal.

The principle being applied is simple: two transmitters, one at each end of the line, feed the signals. The voltage at the line will be the sum of the two signals: U1 + U2. In the actual circuit however this is equivalent to only the half of the sum. At each end, the signal is recovered while the other signal gets rejected. This requires that two identical circuits must be constructed.

The op-amp A1 works as an impedance converter and at the same time functions as the transmitter. The resistor R5 protects the output of A1 from the signals coming from the transmitter at the other end of the line. The sum of the signals is taken from the output of the circuit and fed to the non-inverting input of the differential amplifier composed of the op-amps A2, A3 and A4.

The „right" signal goes through the voltage divider network made of R4, R11 and P1. It is then taken out from the sum of the signals. This „right" signal appears at the OUT output of the circuit.

The resistors must be low tolerance type to achieve the highest same phase suppression possible. This suppression is about 80 dB by 1 kHz and 60 dB by 20 kHz. When using long cables, the suppression can be improved by adjusting C3.

Calibration: Connect the (IN) input to a signal generator and feed a sinus wave of 5 kHz with 1 Veff to the circuit. Attach the twisted pair line (cable) to the (IN/OUT) point of the circuit and short the input of the second circuit at the other end of the cable. Raise the signal frequency to 10 kHz and adjust C3 to get the best results.

65   DX AUDIO FILTER

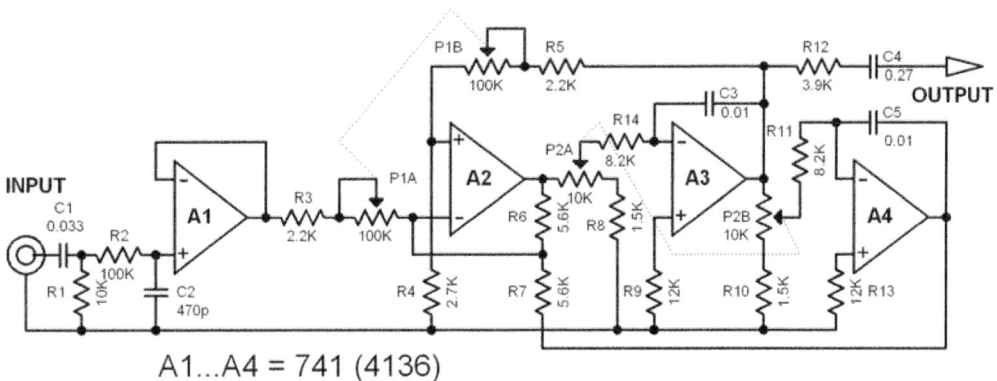

Diagram 65.0 DX Audio Filter

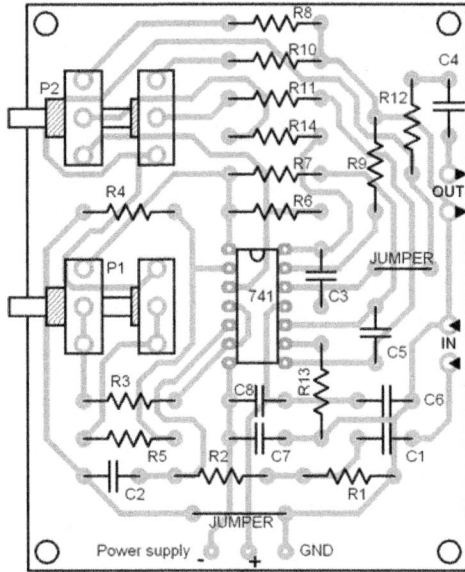

Figure 65.0 Parts Placement Layout for the DX Audio Filter

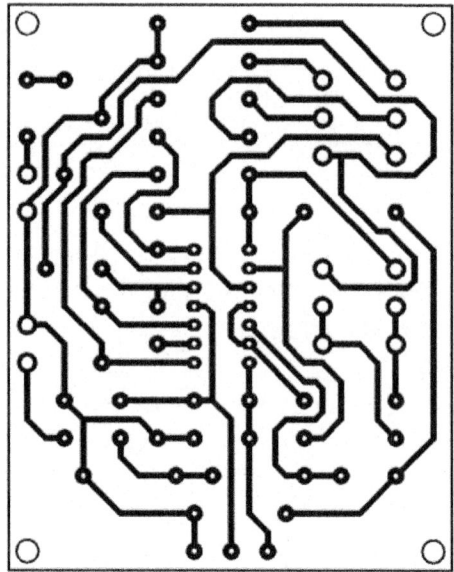

Figure 65.1 Printed Circuit Layout for the DX Audio Filter

Most universal radio receivers (arguably the cheaper ones) have a very wide bandwidth that is not particularly useful for radio amateurs. The better models with narrower bandwidth are almost always priced higher, too high for the reach of an average radio hobbyist.

However, if one wants to use a wide bandwidth receiver to listen to amateur SSB and CW stations with the least amount of interference, this featured electronic circuit can be of great help. This audio band filter cleans out the unwanted noise. It is a state variable type of filter, a bandpass filter with a variable center frequency and bandwidth. If this filter is connected in front of the audio amplifier, one can filter out the unwanted signals from the audio. One requirement for it to work properly is that the filter's center frequency is set exactly to the audio frequency of the wanted signal.

The RC network made of R1,R2,C1 and C2 works as input bandpass filter with a 6 dB cut-off at 500 Hz and 3400 Hz. The op-amp A1 functions as a buffer stage between the input filter and the state variable filter. The state variable filter is composed of the op-amps A2, A3 and A4. The potentiometer adjusts the bandwidth of the filter. The potentiometer P2 adjusts the center frequency of the filter from about 200 Hz up to 2 kHz. By setting these two potentiometers correctly, it is possible to filter out a certain frequency band from the entire audio spectrum.

The power supply is made of two 15 volts. The circuit consumes very little current.

66 MULTI-SOUND SIREN

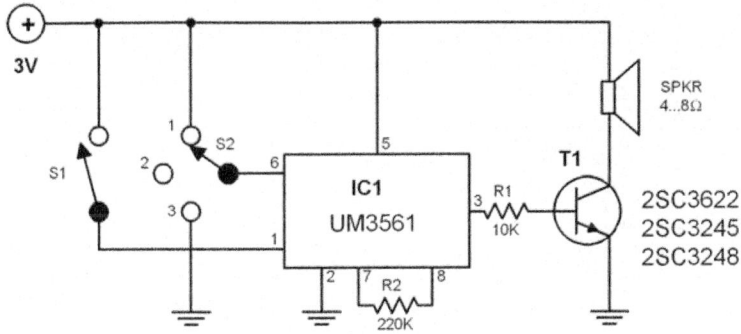

Diagram 66.0   Multi-Sound Siren

This siren can generate four types of siren sounds and is composed of only two active components. The siren sound can be selected through the combination of two switches (see Table 66.0). This circuit is universal and can be applied as a signalling device for alarms, monitors, doorbells, pagers, etc.

S2	S1	SOUND
2	OFF	POLICE
1	OFF	FIRE
3	OFF	AMBULANCE
1	ON	STACCATO

Table 66.0

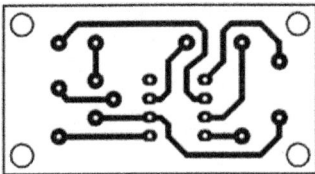

Figure 66.0   Printed Circuit Layout

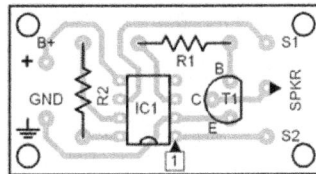

Figure 66.1 Parts Placement Layout

67 IC TREMOLO

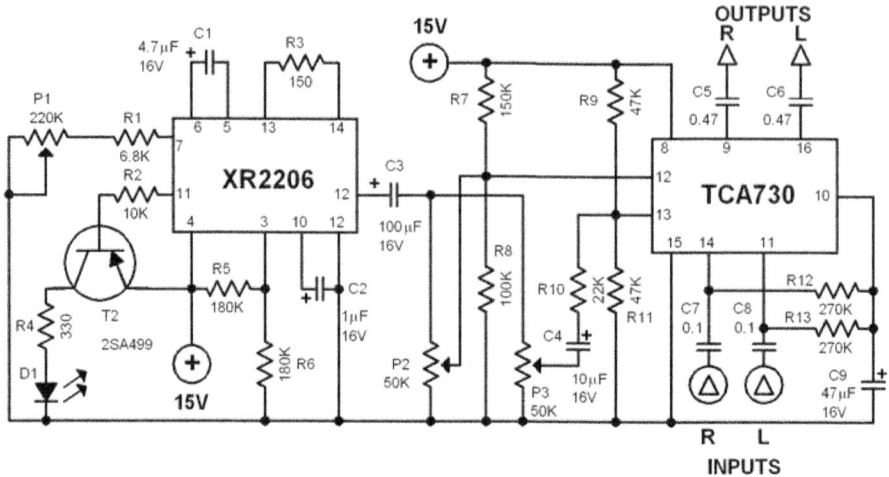

Diagram 67.0 IC Tremolo

Many of the well known tremolo effect circuits have the following disadvantages: the distortions are quite high; the modulation hub and modulation frequency have a narrow bandwidth. The featured circuit allows a modulation hub from 0...100 percent and it is relatively free from distortions. The circuit is useful for stereo channels and it also has the ability to simulate the Lesley effect aka rotating speaker effect.

The circuit uses the TCA730 IC which is designed as an electronic balance and volume regulator with frequency correction. In principle, balance and volume settings are done with a linear potentiometer for both channels. If this potentiometer is replaced with an AC voltage source, a periodic modulation of the input signal can be achieved. This AC voltage source comes from the function generator IC (XR2206). This IC generates square, triangle, and sine wave signals. For this project however only the sine wave is used to create a soft modulation.

The modulation voltage can be varied with P1 from 1 Hz up to 25 Hz. Resistor R3 sets the operational level of the sine wave generator. Resistors R5 and R6 set the DC voltage and the sine wave amplitude at the output. Capacitor C2 is a ripple filter. The square wave output of the XR2206 drives T2 and a LED to optically display the frequency. The modulating voltage reaches pin 13 of TCA730 via P3 and R10. This input functions as the volume control or in this case the volume modulation. The degree of the balance modulation (Lesley effect) can be varied with P2. A regulated power supply using a 7815 IC is recommended. Do not use a non-stabilized power supply since the current variations would influence the modulation negatively. Attach the 7815 IC to a good size heatsink (about 10 cm²).

68 GUITAR SOUND LIMITER

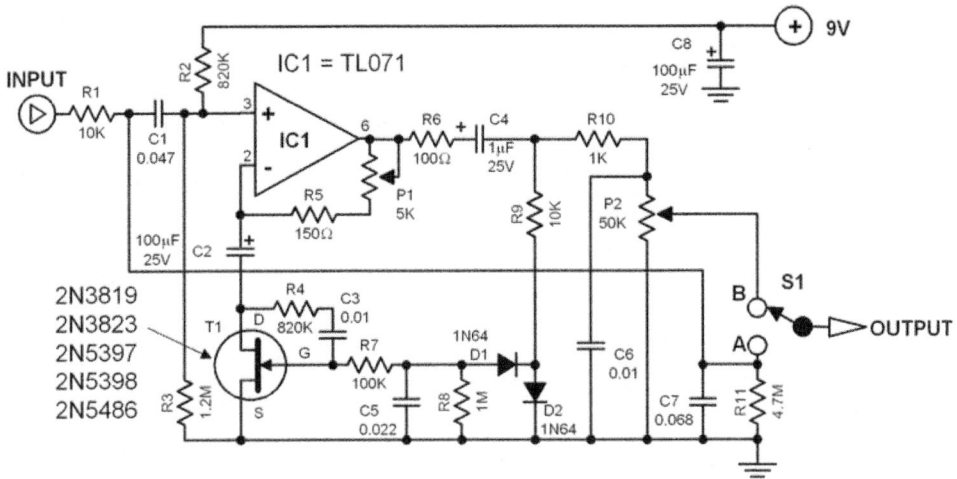

Diagram 68.0   Guitar Sound Limiter

This circuit limits the sound output of an electric guitar by compressing its signal peaks. This limiting technique is usually employed to avoid distortions caused by overdrive.

The electronic circuit is simple. A single IC TL071 works as preamp and the FET transistor works as a voltage controlled resistor. The FET is being controlled by a negative rectified voltage sampled from the output of IC1. When the output increases, the resistance of T1 also increases and reduces the amplification of IC1.

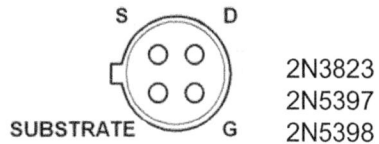

2N3823
2N5397
2N5398

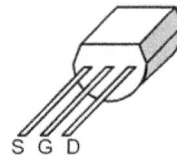

2N3819
2N5486

Calibration: You need an oscilloscope and a signal generator. Feed a 1 kHZ signal of 150 mV into the input and adjust P1 to the maximum amplification without distortion as seen on the oscilloscope. Then, increase the input to 300 mV and adjust P1 back until the distortion is reduced to an acceptable degree.

69 SOUND EFFECTS CIRCUIT

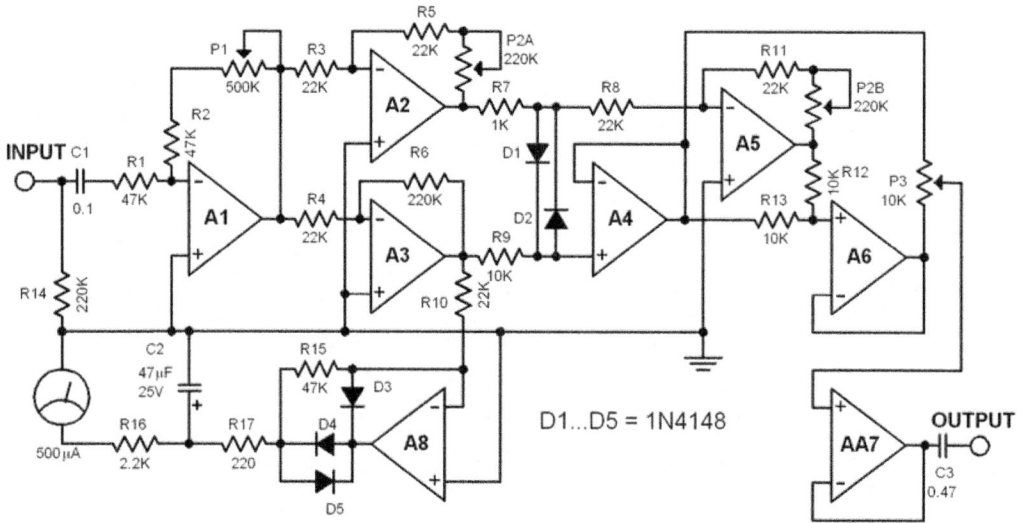

Diagram 69.0 Sound Effects Circuit

This circuit is designed to work as a signal „distorter". If used with an electric guitar, it allows the production of special sound effects. The signal distortion is done by subtracting and mixing the normally clipped input signal with a processed „noise signals". The outcomes are different waveforms. The variations can range from minimal to very strongly clipped signal up to a pulsed waveform which can have increasing dynamics. Potentiometer P3 adjusts the desired waveform while P2 adjusts the degree of distortion. Potentiometer P1 adjusts the input signal level which influences the output waveform.

The gain is dependent on the setting of P1 and the chosen effect degree. It is between 10 dB and 30 dB.

Op-amp A1 is an impedance converter and signal buffer. Opamp A3 amplifies the signal constantly by a factor of 10 (20 dB). Diodes D1 and D2 clip the signal which is then fed to A4. The „noise signal" comes from the junction of R9 and R10. Opamp A7 is another impedance converter. Opamp A8 is used to drive a VU meter.

How to use the sound effect generator: Set the input level with P1 until the VU meter points between 40 and 75% when a guitar string is struck. Select the distortion level with P2 from 0% to 100%. You can now select the waveform with P3.

70 TELEPHONE AMPLIFIER

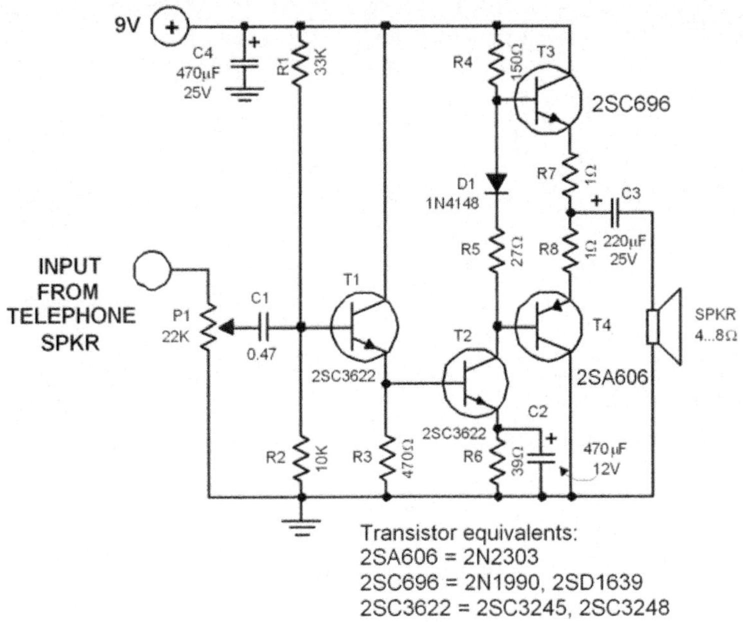

Diagram 70.0 Telephone Amplifier

This house-phone (or telephone) amplifier does not need an inductive pickup. It is directly connected to the phone's speaker capsule. This way, interference is avoided.

The input polarity is not important since the circuit has its own power supply. Potentiometer P1 sets the volume of the audio signal. The standby current consumption is around 30 mA. If you want to get a better sound quality, replace resistor R5 with a 100 ohms trimmer and set the standby current of T3 & T4 to 10 mA.

POWER SUPPLY 9V   BATTERY 9V

HOUSEPHONE
AMPLIFIER

B+

GND

IN

GND

SPKR

SPEAKER 4...8Ω

P1

VOLUME CONTROL

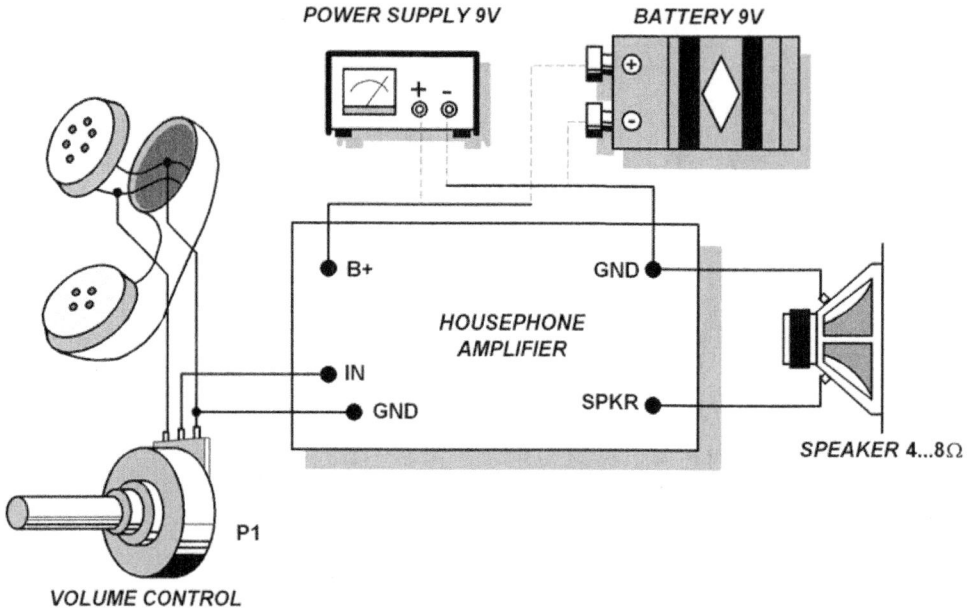

Figure 70.0   External Wiring Layout for Telephone Amplifier

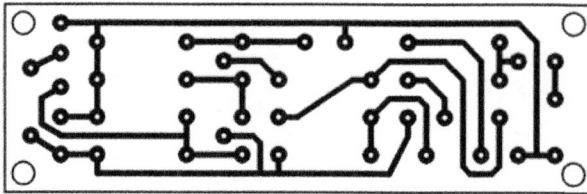

Figure 70.1   Printed Circuit Layout

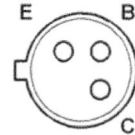

E   B

C

2SA606   2SC696
2N2303   2SD1639
2N1990

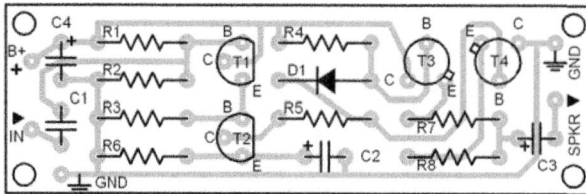

C4
B+
+
C1
IN
R1
R2
R3
R6
T1
T2
D1
R4
R5
R7
R8
C2
T3
T4
C3
B
C
E
B
C
E
B
E
B
E
C
GND
SPKR
GND

Figure 70.2   Parts Placement Layout

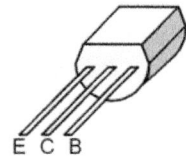

E C B

2SC3622
2SC3245
2SC3248

Transistor equivalents

71 3-CHANNEL AUDIO MIXER

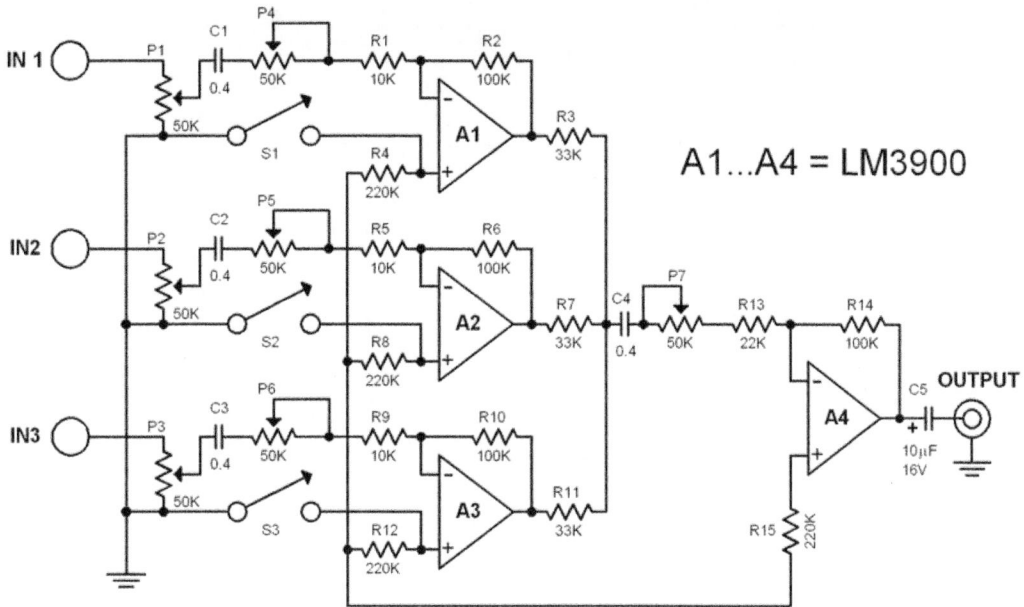

Diagram 71.0 3-Channel Audio Mixer

This mixer circuit uses four Norton amplifiers that are integrated inside the LM3900 IC. The advantage of using this IC is that all four amplifiers need only a single supply. In this circuit, the DC bias is dependent on the feedback since this is a current controlled amplifier. In a Norton amplifier, the output must be set constantly at half of the supply voltage. It allows for a maximum gain without causing distortions. It is called symmetrical limitation at overdrive.

The amplification can be freely chosen. The R2/R1 ratio sets the AC amplification. If R4 is equal to 2R2, then the amplification is null.

Diagram 71.0 shows the complete electronic circuit. The signal level at each input can be varied through P1, P2, and P3. Furthermore, each input can be trimmed to the connected source through the trimmer potentiometers P4, P5, and P6. The resistors at the non-inverting input of the Norton amplifiers set the DC bias and set the output at one half of the the supply voltage. The sum of the signals is then fed to the amplifier A4. The final volume level can be controlled through P7. The switches S1, S2 and S3 either activate or deactivate the input signal.

72   CW AUDIO FILTER

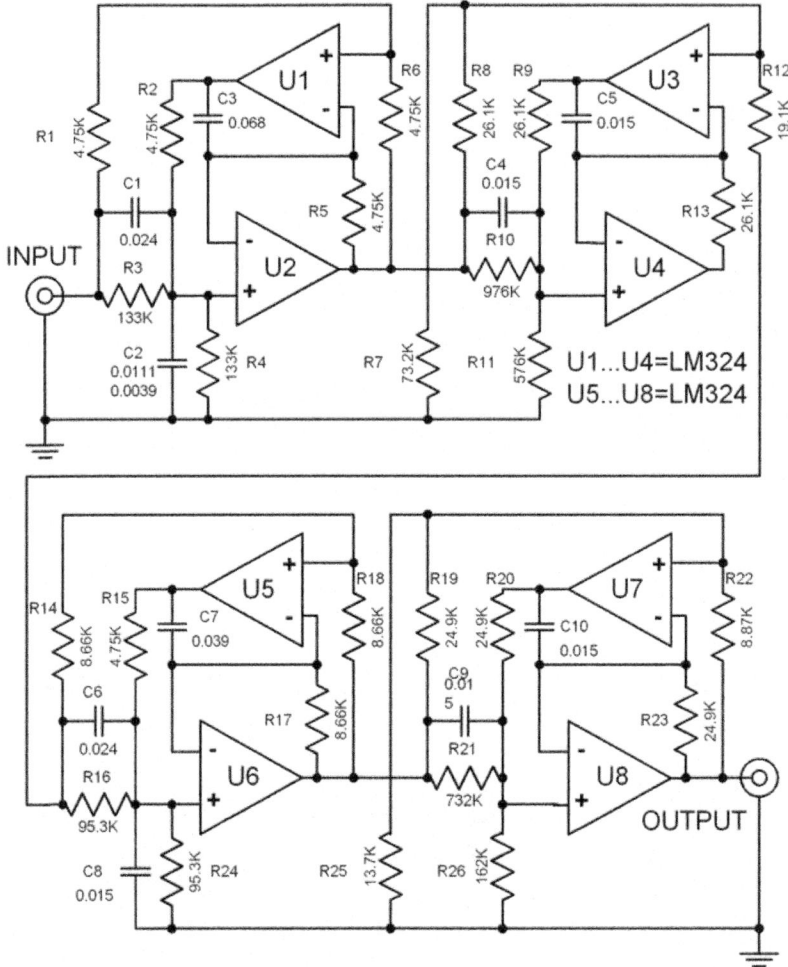

Diagram 72.0 CW Audio Filter

    Morse code telegraphy is a very interesting technique of communication. Many radio amateurs are using this mode because morse telegraphy has a longer range than voice modulation and because morse transmitters are simple and cheap to construct. Morse code communication is however sometimes plague with interference problems. The reception through heavy interference can be very bad specially when simple receivers are used.

The filter circuit featured here is designed to help improve the quality of signal reception. Actually, the filter circuit does not improve the ability of the receiver to receive but instead it filters out unwanted interfering signals from the audio to make the morse signal more readable. The narrow bandpass characteristic of the filter only lets the signals within 380 Hz to 500 Hz pass through. To achieve this level of accuracy, the resistor values must be exactly that what the circuit shows. Sometimes you need to parallel two components in order to get the exact value needed. The circuit needs a symmetrical +15 volts and -15 volts power supply. The filter can be connected directly to the loudspeaker output of the receiver.

73 MUSICAL DOORBELL

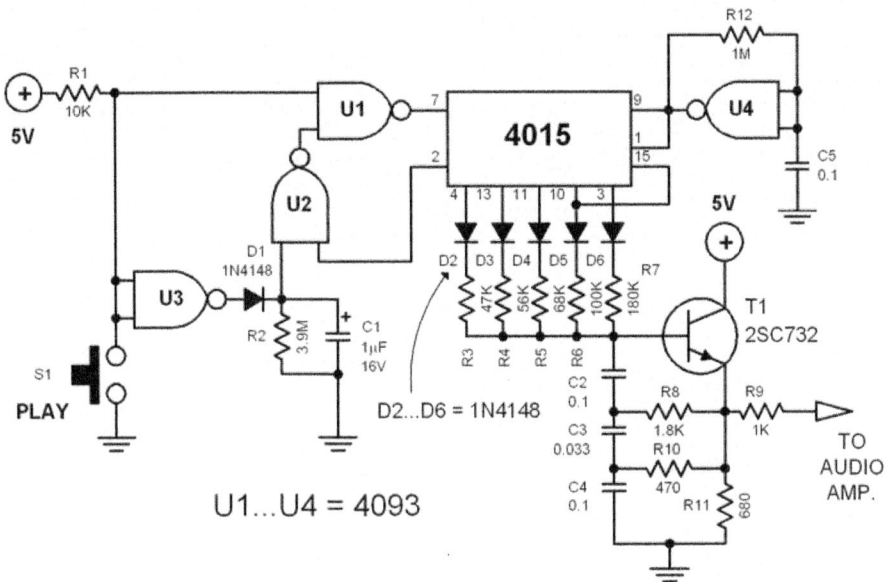

Diagram 73.0 Musical Doorbell

In many cases, a doorbell that sounds off a musical tone is preferable over the common buzz sound. This featured circuit is a musical doorbell. After the button S1 is pressed, a short melody is played. When the button is pressed many times in rapid succession or pressed longer, a different melody is generated and the melody plays longer.

The circuit works this way: when the button S1 is activated the inputs of U3 and one input of U1 switches to logic „0". The data input (pin 7 of IC 4015) becomes logic „1". The 4015 is a static 4-bit shift register. Each clock impulse coming from U4 shifts this logic „1" further in the register. The clock frequency is around 5 Hz.

The number of shifted logic „1" is directly dependent on the length of time the switch S1 is closed. Once at least one shift register is logic „1", a current flows to the base of T1 through a corresponding resistor. The transistor T1 functions as a current controlled oscillator. The tone pitch is dependent on the logic state of the different flip-flop outputs of the shift register. Each clock pulse shifts the logic „1" in the register by one flip-flop. When S1 is pressed one more time at this moment, another logic „1" is added to the register. One output of the register (pin 2) is coupled back to U2 and U3 so that all the logic „1" in the register always run in a loop.

When S1 is released (opened), the register runs until the capacitor C1 gets discharged through R2. When S1 is again pressed (closed), the capacitor C1 stays charged causing the musical bell to play continuously.

The difference between the two ways of activating the switch S1 is that different combinations of logic „1" are inputted into the shift register. These different combinations produce the different melodies played by the circuit. The musical doorbell must be connected to an audio amplifier. The supply voltage is not critical. It can be between 5 and 15 volts. The circuit consumes around 15 milliamperes.

74 VOX SWITCH

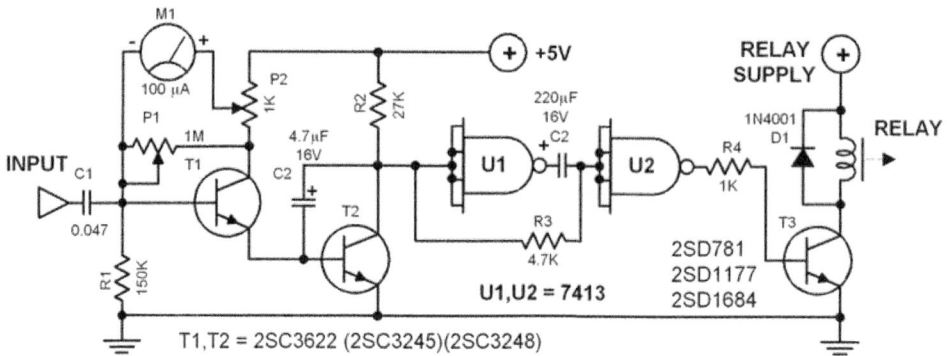

Diagram 74.0 VOX Switch

There are many situations where you need to activate a certain device but you just cannot do it because your hands are busy or just cannot let go. A classic example is operating a radio transceiver while driving a motorcycle. Although you may try to drive and operate the transceiver simultaneously, this will distract and is very dangerous. The best way to solve such problem is to connect a vox (voice operated relay) to the transceiver.

E C B

2SD781
2SD1177
2SD1684

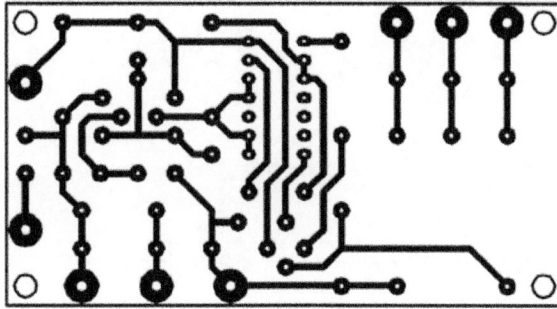

Figure 74.0 Printed Circuit Layout

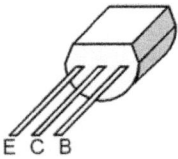

E C B

2SC3622
2SC3245
2SC3248

*Transistor
equivalents*

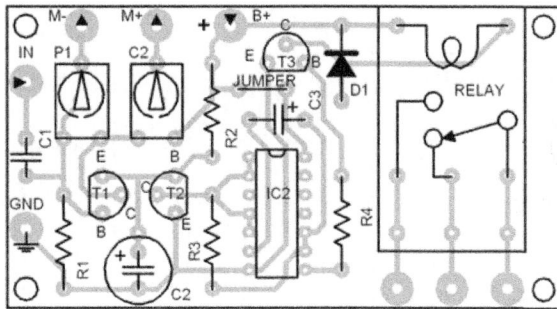

Figure 74.1 Parts Placement Layout

With the vox installed, you can remotely switch the transceiver to transmit mode by just speaking on the mic (preferably built in inside the helmet). One such vox is featured here. This relay circuit is operated by voice. The voice signal may come from a mic amplifier, an audio amplifier, or similar devices. It can be used to automatically control the transmit switch of a transceiver (VOX), a voice operated repeater, or a slide projector film changer. Obviously, the application area is not limited to motorcycling.

The sensitivity can be set through P1. Set it to a level where the vox is immune to the background noise and will only activate with your voice. The analog meter M1 is used in the circuit as a simple audio meter. Potentiometer P2 adjusts the maximum deflection of the meter. You can remove this meter if you want. However if you do so, you have to readjust the P1 to set the desired sensitivity level.

Here's a tip on one practical application of this circuit: this vox can be used as the auto-switch to activate the transmitter module of a vhf repeater system. The vox's input is simply plugged in to the receiver's audio output and its relay terminals are connected to the ptt line of the transmitter module

75 MUSIC SOUND GENERATOR

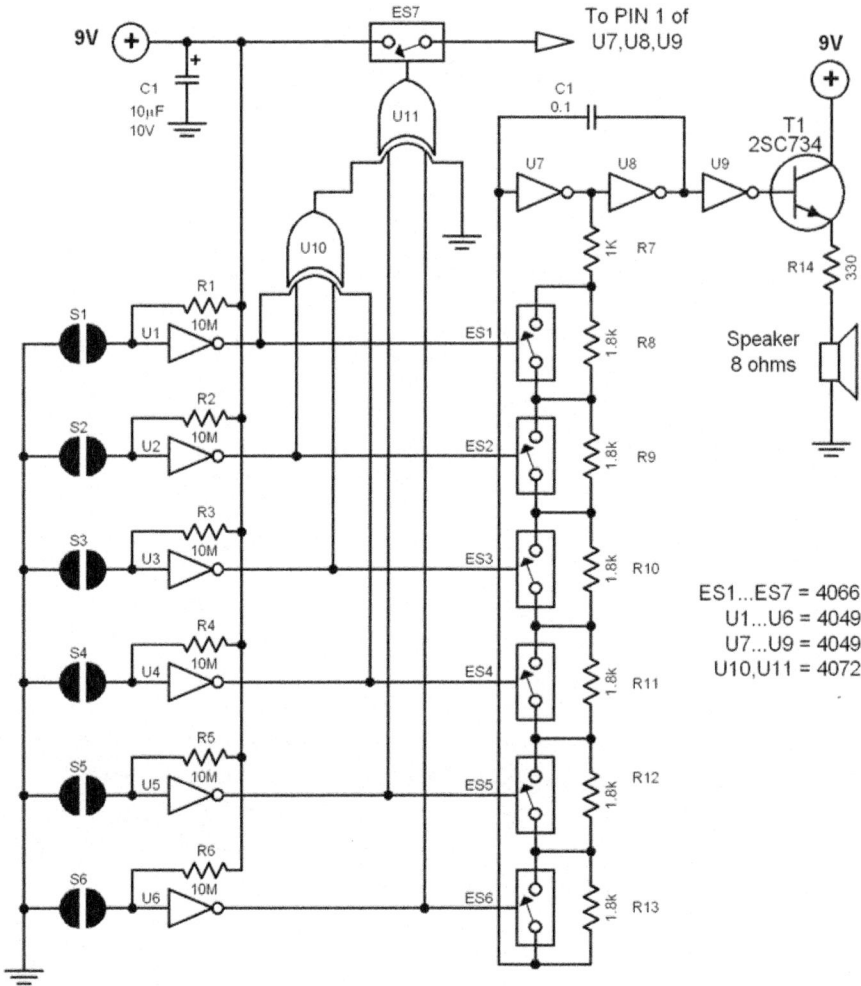

Diagram 75.0 Music Sound Generator

This electronic circuit generates a tone when the buttons (S1 ... S6) are touched. The tone frequency is dependent on the button that is being touched.

76 REGULATED MIC PREAMP

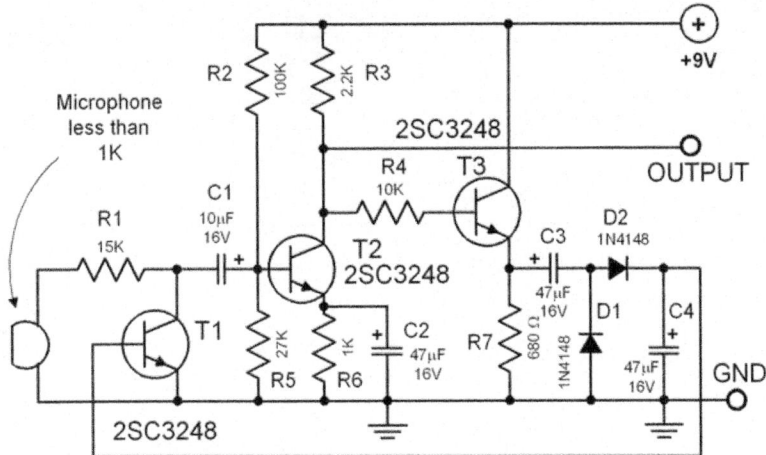

Transistor equivalents:
2SC3248 = 2SC3622, 2SC3245, 2SC2459, 2SC3112, 2SC2675

Diagram 76.0 Regulated Mic Preamp

This microphone preamp boosts weak audio signals before they are introduced into another circuit for further processing. This circuit is designed to be used particularly in modulating radio transmitters.

The audio signal picked up by the microphone is first amplified by T2 and passed on to T3 which functions as emitter follower. The signal is then rectified by diodes D2-D1 and smoothed by capacitor C4. This signal is then used as feedback to control T1. When the audio signal is very strong, T1 conducts more current thereby reducing the signal that reaches T2. The maximum input voltage is about 1 Vp-p. The microphone can also be replaced by a speaker.

Parts List:

Resistors:
R1=   15K
R2=   100K
R3=   22K
R4=   10K
R5=   27K
R6=   1K
R7=   680Ω

Capacitors:
C1=   10µF/16V electrolytic
C2,C3,C4=   47µF/16V electrolytic

Transistors:
T1,T2,T3= 2SC3248

77   SIREN

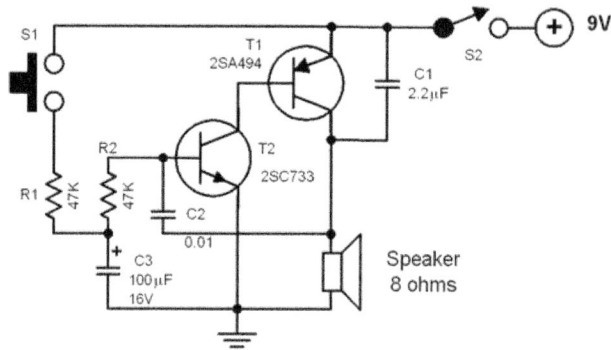

Diagram 77.0 Siren

This circuit generates a tone that sounds very similar to a siren. The generator part of the circuit is made of the combination of PNP and NPN transistors. Together, the two transistors build up a free running multivibrator. If the C2 capacitor was connected to the positive line of the power supply, it would have worked as a constant frequency oscillator.

However, we don't want a constant frequency oscillator. We want a siren. So to generate an up and down going signal tone, the resistor R2 is fed from an RC circuit (made of R1 and C1). When the switch S1 is pressed, the capacitor C1 charges via R1 slowly until it reaches the maximum voltage level of 4 volts. This increasing voltage results to a decreasing time constant at the R2/C2 junction. This furthermore results to an increasing frequency of the multivibrator. After the switch S1 is released, the capacitor C1 discharges slowly resulting to a decreasing frequency cycle. Through the combination of the two time constants a sawtooth waveform is generated.

The signal coming from the speaker will be an increasing or decreasing tone depending on whether the switch S1 is pressed or released.

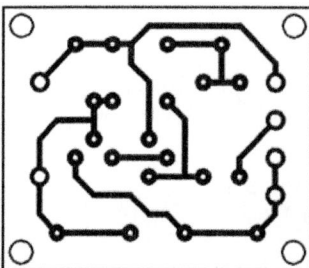

Figure 77.0
Printed Circuit Layout

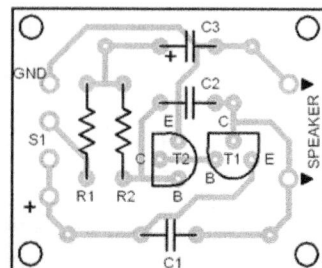

Figure 77.1
Parts Placement Layout

78 GUITAR SOUND LIMITER

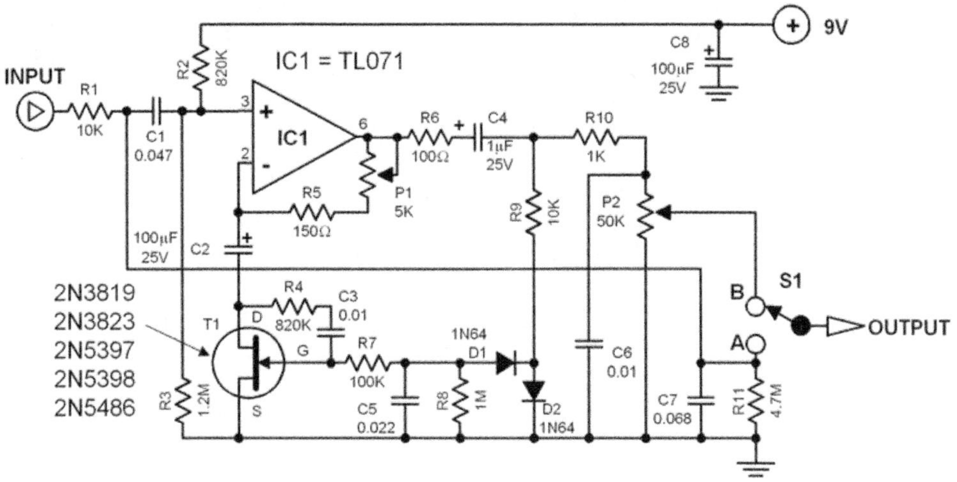

Diagram 78.0 Guitar Sound Limiter

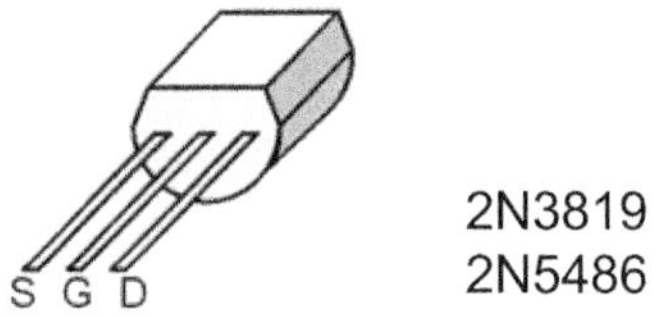

Transistor equivalents

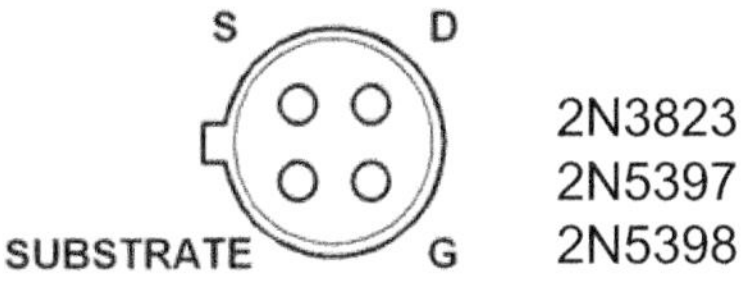

This circuit limits the sound output of an electric guitar by compressing its signal peaks. This limiting technique is usually employed to avoid distortions caused by overdrive.

The circuit is simple. A single IC TL071 works as preamp and the FET transistor works as a voltage controlled resistor. The FET is being controlled by a negative rectified voltage sampled from the output of IC1. When the output increases, the resistance of T1 also increases and reduces the amplification of IC1.

Calibration: You need an oscilloscope and a signal generator. A 1 kHZ signal of 150 mV is fed into the input and P1 is adjusted to the maximum amplification without distortion as seen in the oscilloscope. Then, the input is increased to 300 mV and P1 is adjusted back until the distortion is reduced to an acceptable degree.

79 WHISTLE PROCESSOR

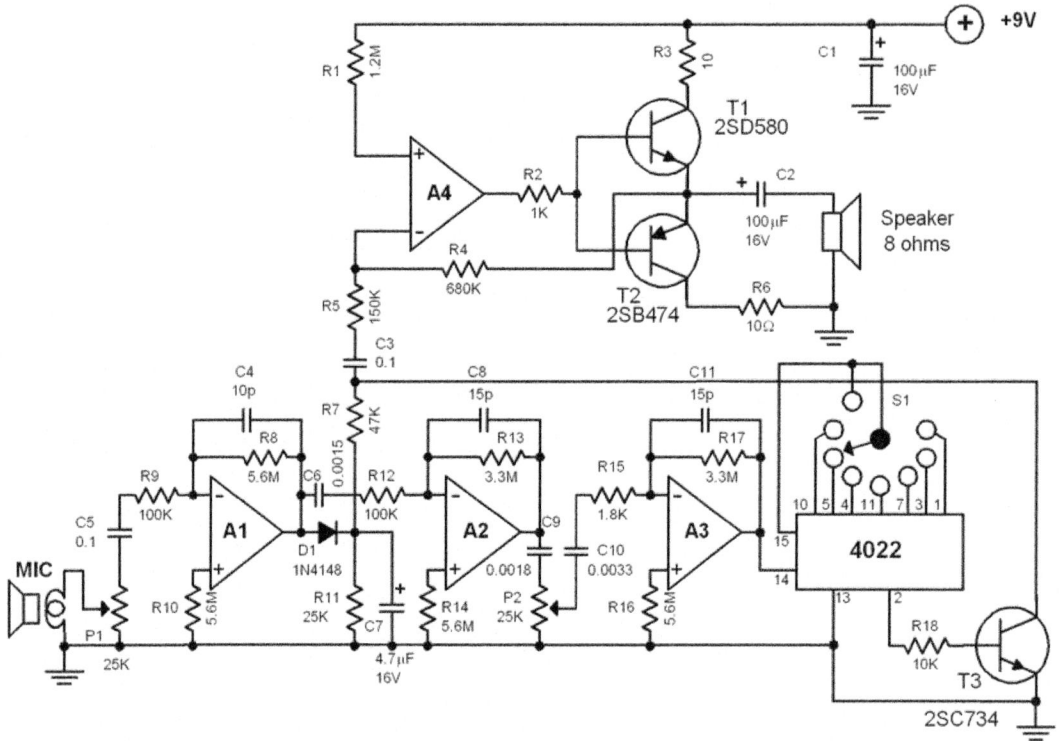

Diagram 79.0 Whistle Processor

This circuit processes the whistle into a tone with lower frequency but with almost the same amplitude as the original tone. The resulting tone is lower but with the same dynamics. The microphone being used is of the crystal type. The switch S1 selects the point where the counter will be reset. This way, the divider factor from 1 up to 8 can be selected.

The circuit can be used as an octave divider with an electric guitar.

80 SOUND GENERATOR

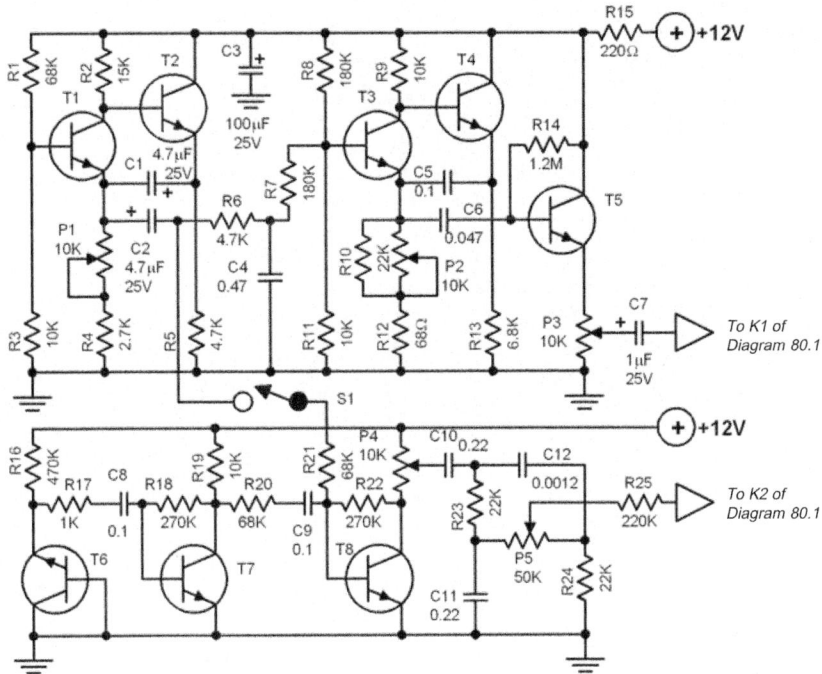

Diagram 80.0 Sound Generator

The construction of this seemingly complicated electronic circuit will be awarded with the joy of generating almost unlimited sound effects. The circuit can produce different sounds like the heavy chopping of a helicopter´s blades up to the fine chirping of a bird.

A vibrato generator (T1 and T2) modulates a tone generator (T3 and T4). The frequency of the vibrato generator can be varied through P2. A second vibrato generator is also built (T6...T10) and has the same function as the first one described.

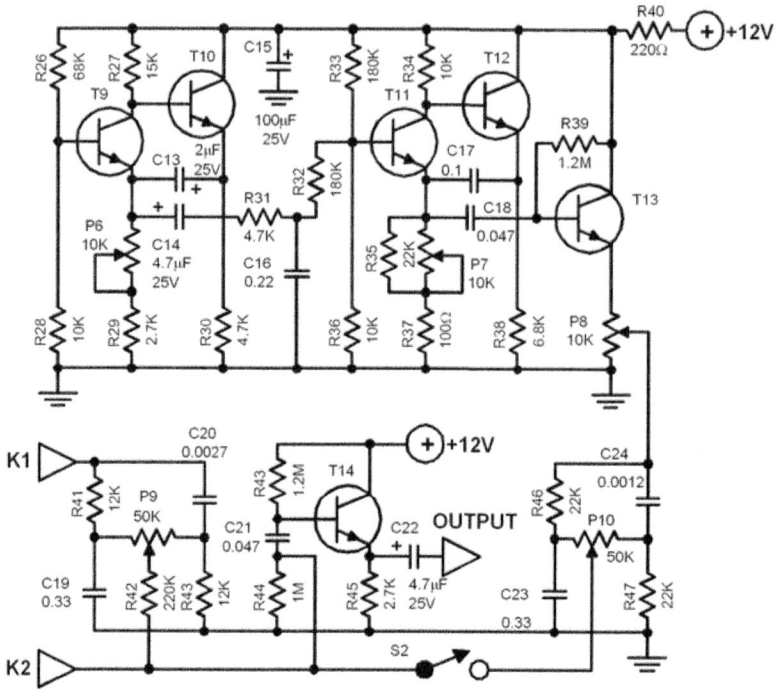

Diagram 80.1 Sound Generator (part 2)

The noise sound is produced by T6 while transistors T7 and T8 amplify the generated noise sound. The potentiometers P3, P8 and P4 influence the amplitude of the generated signal and create the desired sound effect. Potentiometers P5, P9, and P10 produce the desired noise envelope and strength. The mixed tone and noise signals appear at the emitter of T14. Experimenting with different potentiometer settings will certainly produce hours of enjoyment and lots of unique sound effects.

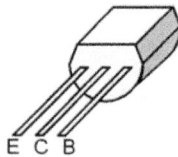

E C B

2SC3622
2SC3245
2SC3249

81 AUTOMATIC VOLUME CONTROL

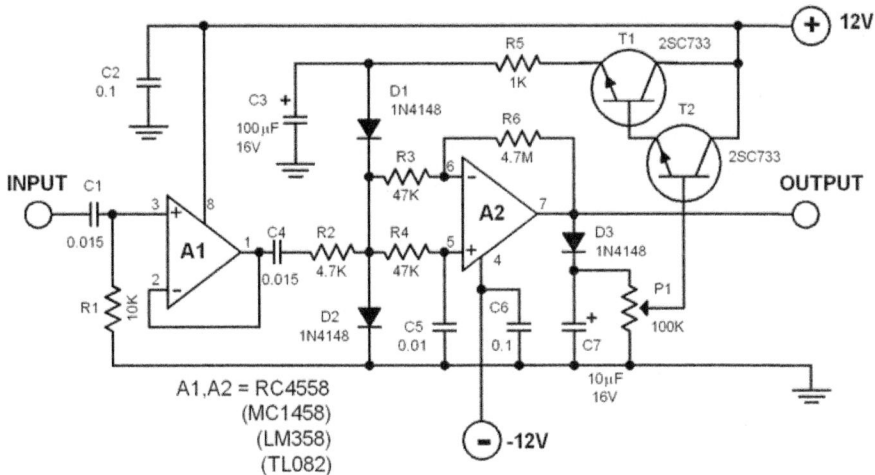

Diagram 81.0 Automatic Volume Control

The featured circuit controls a volume line automatically. It delivers an output voltage of approximately 4 volts peak to peak. This voltage remains relatively constant by input voltages ranging from several hundred millivolts to several volts. This electronic circuit is not highly recommended for HIFI applications since its noise factor is too much above the accepted level. However, it is very useful in tape recording of computer programs. The constant amplitude is desirable in such applications.

The op-amp A1 works as an input signal buffer. If we remove the diodes D1 and D2, this op-amp will work as an amplifier. The DC bias of A1 is done over R4 and C5. This little trick allows the A2 to limit the DC level at its input to a maximum amplification of 100 times. The offset bias is relatively constant. The amplified signal from A2 is rectified and filtered by D3 and C7. A sample of this rectified signal is fed to the regulator diodes D1 and D2 over the transistors T1 and T2. Trimmer P1 controls this sample signal. The higher this signal is, the higher is the current flowing through the diodes.

These regulator diodes have a non-linear curve. Their resistance decrease with increasing current. The input signal gets grounded more or less through the diodes. To put it in another way: the diodes work as an attenuator with increasing effect by increasing current through the diodes.

82 ELECTRONIC POTENTIOMETER

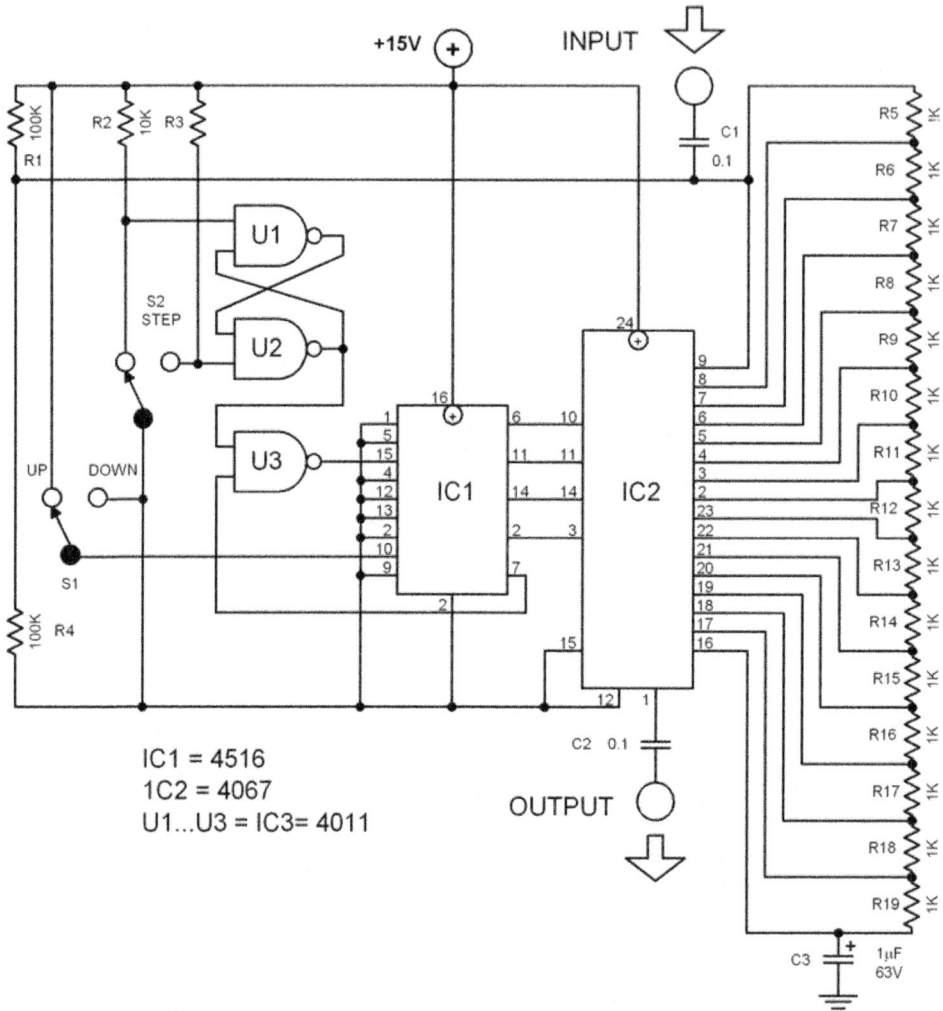

Diagram 82.0   Electronic Potentiometer

IC1 = 4516
1C2 = 4067
U1...U3 = IC3= 4011

This is a digitally controllable potentiometer. The heart of the circuit is a 16 channel analog multiplexer. Controlled by the BCD value at the inputs, the output pin of the multiplexer (pin 1) is connected to any one of the 16 outputs. The output pins are connected with 1k resistors. The IC2 can therefore be considered as a linear potentiometer with 16 steps. The total resistance of this potentiometer is 15k. Other values or characteristics like positive logarithmic can also be achieved by using different resistor values. The setting of the potentiometer is controlled by the counter IC 4516. The position of S1 determines whether the resistance increases or decreases every time S2 is pressed.

83 SOUND EFFECTS CIRCUIT

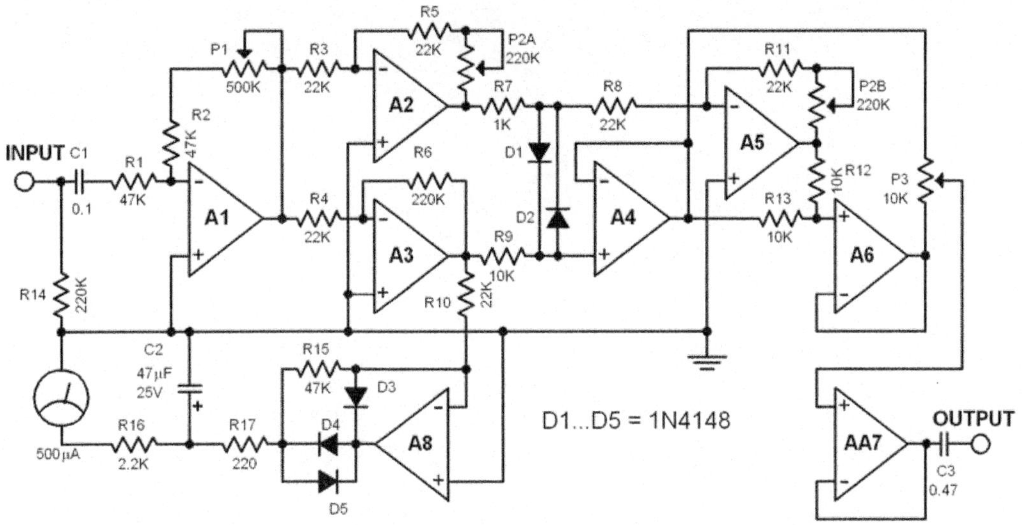

Diagram 83.0 Sound Effects Circuit

This circuit is designed to work as a signal „distorter". If used with an electric guitar, it allows the production of special sound effects. The signal distortion is done by subtracting and mixing the normally clipped input signal with a processed „noise signals". The outcomes are different waveforms. The variations can range from minimal to very strongly clipped signal up to a pulsed waveform which can have increasing dynamics. Potentiometer P3 adjusts the desired waveform while P2 adjusts the degree of distortion. Potentiometer P1 adjusts the input signal level which influences the output waveform.

The gain is dependent on the setting of P1 and the chosen effect degree. It is between 10 dB and 30 dB.

Op-amp A1 is an impedance converter and signal buffer. Opamp A3 amplifies the signal constantly by a factor of 10 (20 dB). Diodes D1 and D2 clip the signal which is then fed to A4. The „noise signal" comes from the junction of R9 and R10. Opamp A7 is another impedance converter. Opamp A8 is used to drive a VU meter.

How to use the sound effect generator: Set the input level with P1 until the VU meter points between 40 and 75% when a guitar string is struck. Select the distortion level with P2 from 0% to 100%. You can now select the waveform with P3.

84 HEADPHONE AMPLIFIER

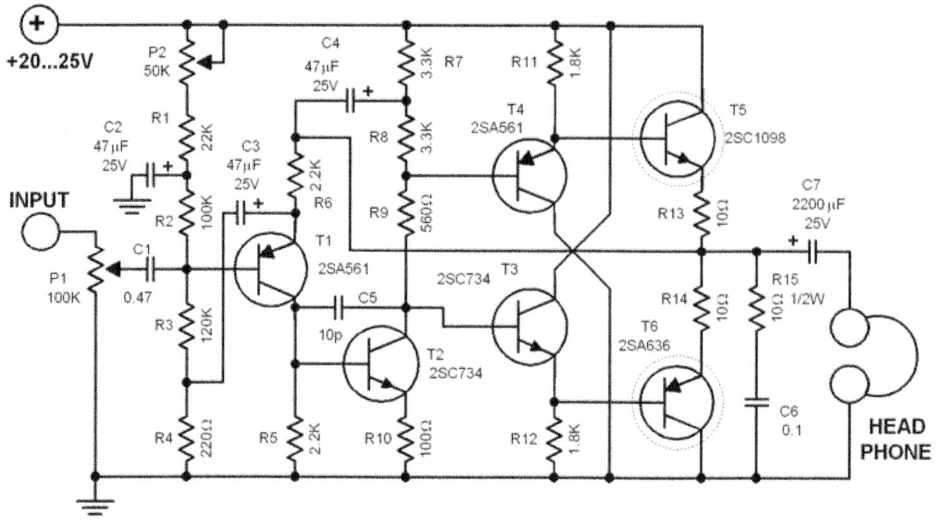

Diagram 84.0 Headphone Amplifier

Typically, a headphone is connected to the speaker output of the final amplifier stages through a voltage divider circuit. However, this simple design has two distinct disadvantages. Firstly, the headphone volume cannot be varied independently from the main speaker when the main speaker is switched on at the same time. Secondly, the voltage divider circuit causes attenuation and at the same time affects the bass output negatively.

The solution to the problem is an independent amplifier for the headphones such as the circuit shown in diagram 84.0. This amplifier is connected to the output of the final amplifier through the potentiometer P1. If a stereo headphone is used, this potentiometer must be replaced with a stereo type. Furthermore, the entire circuit must be duplicated for the second channel.

The headphone amplifier delivers an output of around 1 watt. Use a power supply rated at 350 mA. The amplifier gain is dependent on the resistors R4 and R6. The values shown in the circuit gives a gain of 11. The voltage at the junction of R13 and R14 must be set at 50% of the power supply. This can be set through P2. The standby current through the final transistors is about 50 ... 110 mA.

This page intentionally left blank

This page intentionally left blank

APPENDICES

Specifications of the transistors used in the projects

Descriptive Part of the Table:

Type
The original type designation has been taken over directly from the manufacturers, with the abbreviation of the manufacturer added in brackets only in those cases in which different manufacturers used the same type designation.

Mat.
The materials used are abbreviated as follows:

Ge	Germanium
Mos	MOS technology (metal oxide silicon)
Si	Silicon
V-MOS	Vertical MOS technology

Pol.
The polarities used are abbreviated as follows:

npn	NPN structure
n-ch	N channel type (FET)
n-p	More than one transistor with different polarities in one case
pnp	PNP structure
p-ch	P channel type (FET)

Abbreviations used in the following table:

A	Antenna amplifer		FET	Field-effect transistor
AGC	Regulating steps		FET-depl.	Field-effecttransistor, depletion type
AF	AF range		FET-enh.	Field-effect transistor, enhancement type
AM	AM range		FM	FM range
CATV	Broad band cable amplifier		fs	Fast switch
CB	CB-radio		HD	Horizontal deflection
CTV	Colour television application		hi-rel	high reliability
chop	Chopper		Idss	Drain source short-circuit current (FET)
Darl	Darlington transistor		IF	IF applications
dg	Dual Gate (FET)		in	Input stages
double	Paired types		iso	insulated
dr	Driver stages		In	Low noise
dual	Dual transistor (differential amplifier)		min	Miniaturised version
			mix	Mixer stages
end	Final stages		nixie	Digital display tube

osc	Oscillator stages	Ugs	Gate source voltage
pow	Power stages	UHF	UHF range > 250MHz
radiation	Aerospace applications	uni	Universal type
	(radiation-proof)	Up	Pinch-off voltage (FET)
RF	RF range	VD	Vertical deflection
s	Switch	VHF	VHF range 100-250
SMP	Switch-mode power supply	MHz	
SSB	Single side-band operation	Vid	Video output stages
Stabi	Stabilisation	+Diode, +di	With integrated diode
sym	Symmetrical types	../..ns	turn-on/turn-off time
TV	Television applications		

Data Part of the Table:

In the case of the ratings, either average values are quoted (< = max.) or lower (> = min.) guaranteed values. As a rule apply at 25°C, unless otherwise indicated.

Uc

With transistors, the usual situation is for U_{CBO}(colletor base reverse bias) to be quoted, or U_{CEO} and U_{CEO} (collector emitter reverse bias). With FETs, U_{DS} (drain source voltage) is always quoted.

Ic

With transistors, I_c (collector current) is always quoted. If this is followed by (ss) in brackets, I_{CM} is quoted, i.e. the peak value of the collector current. With FETs, I_D (drain current) is always quoted.

$Ptot$

As a rule, the total leakage power Ptot is quoted, with RF types we always quote the RF output power P_Q, with corresponding frequency in brackets.

Amplification

The DC current gain B(h_{FE}) or the short-circuit current gain ß(h_{fe}) are always quoted as guaranteed values.

f_T

The transition frequency is always qouted in MHz.

Specifications of the transistors used in the projects

Type	Mat.	Pol.	Description	UC [Vmax]	IC [Amax]	Ptot [Wmax]	Current Gain	fT [MHz]
MJ3001	Si	npn	Darl+diode,pow	60	10.00	150.00($25°C)	>10	
MJE243	Si	npn	AF-s-pow	100	4.00	1.50($25°C)	40.120	>40.00
MJE244	Si	npn	AF-s-pow	100	4.00	1.50($25°C)	>25	>40.00
MJE253	Si	npn	AF-s-pow	100	4.00	1.50($25°C)	40-120	>40.00
MJE4350	Si	pnp	AF-end,s-pow	100	16.00	125.00($25°C)	15	>1.00
MJE5170	Si	pnp	uni-pow	120	6.00	2.00($25°C)	15-100	>1.00
MJE5180	Si	npn	uni-pow	120	6.00	2.00($25°C)	15-100	>1.00
MPF102	Si	n-ch	FET,VHF-in,sym,mix 25V, Idss>2mA,Up<V					
MPF106	Si	n-ch	FET,VHF 25V,Idss>4mA,Up<8V					
MPS-A29	Si	npn	Darl	100	0.50	1.50($25°C)	>10	>125.00
2N708	Si	npn	s	40/15	0.20	1.20(25°C)	>15	480.00
2N1711	Si	npn	uni	75	0.50	3.00(25°C)	75	>70.00
2N1889	Si	npn	AF-s	100/60	0.50	3.00(25°C)	40-120	>50.00
2N1890	Si	npn	AF-s	100/60	0.50	3.00(25°C)	100-300	>60.00
2N1990	Si	npn	nixie	100	1.00	2.00(25°C)	>25	
2N2102	Si	npn	AF-s	120/65	1.00	5.00(25°C)	40-120	>120.00
2N2222	Si	npn	ini	0		1.80(25°C)		
2N2368	Si	npn	fs	40/15	0.20	1.20(25°C)	20-60	>400.00
2N2369	Si	npn	fs	40/15	0.20	1.20(25°C)	40-120	>500.00
2N2905	Si	pnp	uni	60/40	0.60	3.00(25°C)	100-300	>200.00
2N2904	Si	pnp	uni	60/40	0.60	3.00(25°C)	40-120	>200.00
2N3019	Si	npn	uni	140/80	1.00	5.00(25°C)	100-300	>100.00
2N3020	Si	npn	uni	140/80	1.00	5.00(25°C)	40-120	>80.00
2N3055	Si	npn	AF-s-pow	100/60	15.00	115.00($25°C)	20-70	>2.50
2N3109	Si	npn	AF-s	80/40	1.00	5.00(25°C)	100-300	>70.00
2N3110	Si	npn	AF-s	80/40	1.00	5.00(25°C)	40-120	>60.00
2N3367	Si	n-ch	FET,uni,In	40V,Idss>0.5mA,Up<2.5V				
2N3370	Si	n-ch	FET,uni,In	40V,Idss>0.1mA,Up3.2V				
2N3454	Si	n-ch	FET,uni	50V,Idss>0.05mA,Up<2.3V				
2N3819	Si	n-ch	FET,VHF,uni,sym	25V,Idss>2mA,Up<8V				
2N3823	Si	n-ch	FET,VHF,In	30V,Idss>4mA,Up<8V				
2N3903	Si	npn	uni	60/40	0.20	1.50(25°C)	50-150	>250.00
2N3904	Si	npn	uni	60/40	0.20	1.50(25°C)	100-300	>300.00
2N3905	Si	pnp	uni	40	0.20	1.50(25°C)	50-150	>200.0
2N3906	Si	pnp	uni	40	0.20	1.50(25°C)	100-300	>250.00
2N4118	Si	n-ch	FET,uni	40V,Idss>0.08mA,Up<3V				
2N5294	Si	npn	AF-s-pow	80/70	4.00	1.80($25°C)	30-120	>0.80
2N5397	Si	n-ch	FET,VHF/UHF	25V,Idss>10mA,Up<6V				
2N5398	Si	n-ch	FET,VHF/UHF	25V,Idss>5mA,Up<6V				
2N5486	Si	n-ch	FET,VHF/UHF	25V,Idss>8mA,Up<6V				
2N6038	Si	npn	Darl+diode,pow	60	4.00	1.50($25°C)	>10	>25.00
2N6039	Si	npn	Darl+diode,pow	80	4.00	1.50($25°C)	>10	>25.00
2N6283	Si	npn	Darl+diode,pow	80	20.00	160.00($25°C)	>10	>4.00
2N6284	Si	npn	Darl+diode,pow	100	20.00	160.0($25°C)	>10	>4.00
2N6412	Si	npn	AF-s-pow	60/40	4.00	15.00($25°C)	>5	>50.00
2N6414	Si	pnp	AF-s-pow	80/60	4.00	15.00($25°C)	>5	>50.00
2SA511	Si	pnp	AF/RF/s	90/80	1.50	8.00(25°)	30-150	60.00
2SA597	Si	pnp	RF-s	50/40	1.00	6.00($25°C)	10-250	400.00
2SA761	Si	pnp	uni	110	2.00	6.30($25°)	50-240	80.00
2SA970	Si	pnp	AF,In	120	0.10	0.30(25°C)	200-700	100.0

Specifications of the transistors used in the projects

Type	Mat.	Pol.	Description	UC [Vmax]	IC [Amax]	Ptot [Wmax]	Current Gain	fT [MHz]
2SA1016	Si	pnp	uni,ln	120/100	0.05	0.40(25°)	160-960	110.00
2SA1123	Si	pnp	uni,ln	150	0.05	0.7(25°)	65-450	200.00
2SA1136	Si	pnp	AF-in,ln	120/100	0.10	0.30(25°C)	120-560	90.00
2SA1137	Si	pnp	AF-in,on	80	0.10	0.30(25°C)	120-560	90.00
2SA1141	Si	pnp	AF/Rf-pow	115	10.00	2.00($25°C)	100	80.00
2SA1285	Si	pnp	uni	120	0.20	0.90(25°C)	150-800	200.00
2SA1285A	Si	pnp	uni	150	0.10	0.90(25°C)	150-500	200.00
2SA1515	Si	pnp	uni	40/32	1.00	0.50(25°C)	82-390	150.00
2SA1705	Si	pnp	AF,s	60/50	1.00	0.90(25°C)	>30	150.00
2SA1706	Si	pnp	AF-s	60/50	2.00	1.00(25°C)	>40	150.00
2SB633	Si	pnp	AF-s-pow	100/85	6.00	40.00($25°C)	40-320	15.00
2SB764	Si	pnp	uni	60/50	1.00	0.90(25°C)	60-320	150.00
2SB822	Si	pnp	Af-dr/end	40/32	2.00	0.75(25°C)	82-390	100.00
2SB826	Si	pnp	s-pow	60/50	7.00	60.00($25°C)	>30	10.00
2SB867	Si	pnp	AF/s-pow,lo-sat	130/80	3.00	30.00($25°C)	60-260	30.00
2SB868	Si	pnp	AF/s-pow,lo-sat	130/80	4.00	35.00($25°C)	60-260	30.00
2SB869	Si	pnp	AF/s-pow,lo-sat	130/80	5.00	40.00($25°C)	60-260	30.00
2SB870	Si	pnp	AF/s-pow,lo-sat	120/80	7.00	40.00($25°C)	60-260	30.00
2SB874	Si	pnp	AF/s-pow, TV-VD	100/60	2.00	20.00($25°C)	>40	250.00
2SB909	Si	pnp	AF-dr/end	40/32	1.00	1.00(25°C)	82-390	150.00
2SB911	Si	pnp	AF-dr/end	40/32	2.00	1.00(25°C)	82-390	100.0
2SB920	Si	pnp		120/80				
2SB921	Si	pnp		120/80				
2SB1064	Si	pnp	AF-s-pow	60/50	3.00	1.50($25°)	60-320	70.00
2SB1114	Si	pnp	min,uni	20	2.00	2.00($25°C)	135-600	180.00
2SB1116	Si	pnp	uni	60/50	1.00	0.75(25°C)	135-600	120.00
2SB1142	Si	pnp	s-pow	60/50	2.50	10.00(25°C)	>35	140.00
2SB1143	Si	pnp	s-pow	60/50	4.00	10.00(25°C)	>40	150.00
2SB1144	Si	pnp	AF/s-pow,lo-sat	120/100	1.50	10.00(25°C)	>30	100.00
2SB1230	Si	pnp	AF/s-pow,lo-sat	110/100	15.00	100.00($25°C)	50-140	
2SB1231	Si	pnp	AF/s-pow,lo-sat	110/100	25.00	120.00($25°C)	50-140	
2SB1232	Si	pnp	AF/s-pow,lo-sat	110/100	40.00	150.00($25°C)	50-140	
2SC270	Si	npn	s-pow	270/75	5.00	50.00($25°C)	24-40	22.00
2SC460	Si	npn	AM-in/mix/osc	30	0.10	0.20(25°C)	35-200	230.00
2SC696	Si	npn	uni	100/60	3.00	0.75(25°C)	30-173	100.00
2SC763	Si	npn	VHF	25/12	0.02	0.10(25°C)	20-300	>400.00
2SC829	Si	npn	AM/FM-in/mix/osc	30/20	0.03	0.40(25°C)	40-500	230.00
2SC959	Si	npn	uni	120/80	0.70	0.70(25°C)	40-200	100.00
2SC1324	Si	npn	UHF-CATV	35/25	0.15	3.00(25°C)	10-35	
2SC1876	Si	npn	Darl	100/70	0.50	0.80(25°C)	>20	
2SC2124	Si	npn	TV-HD	220/800	2.00	5.00($90°C)	20	4.00
2SC2125	Si	npn	TV-HD	220/800	5.00	50.00($25°C)	8-25	5.00
2SC2270	Si	npn	lo-sat	50/20	5.00	1.00($25°C)	>70	100.00
2SC2334	Si	npn	s-pow,dc-dc conv.	150/100	7.00	40.00($25°C)	>20	
2SC2459	Si	npn	uni	120	0.10	0.20(25°C)	200-700	100.00
2SC2675	Si	npn	AF,ln	80	0.10	0.30(25°C)	180-820	120.00
2SC2724	Si	npn	FM-IF	30/25	0.03	0.20(25°C)	25-300	200.00
2SC3112	Si	npn	AF,ln	50	0.15	0.40(25°C)	600-3600	250.00
2SC3179	Si	npn	AF-pow	80/60	4.00	30.00($25°C)	100	15.00
2SC3245	Si	npn	uni	120	0.10	0.90(25°C)	150-800	200.00

Specifications of the transistors used in the projects

Type	Mat.	Pol.	Description	UC [Vmax]	IC [Amax]	Ptot [Wmax]	Current Gain	fT [MHz]
2SC3245A	Si	npn	uni	150	0.10	0.90(25°C)	400-800	200.00
2SC3248	Si	npn	uni	180	0.10	0.90(25°C)	150	130.00
2SC3358	Si	npn	UHF	20/12	0.10	0.25(25°C)	50-300	7000.00
2SC3420	Si	npn	lo-sat	50/20	5.00	10.00(25°C)	>70	100.00
2SC3622	Si	npn	AF-s,hi-beta	60/50	0.15	0.25(25°C)	1000-3200	250.00
2SC4308	Si	npn	VHF-A	30/20	0.30	0.60(25°C)	50-200	2500.00
2SD386	Si	npn	TV-VD	200/120	3.00	1.75($25°C)	40-320	8.00
2SD406	Si	npn	Darl	100	2.00	15.00(25°C)	>2000	
2SD613	Si	npn	AF-s-pow	100/85	6.00	40.00($25°C)	40-320	15.00
2SD614	Si	npn	Darl	100/80	3.00	0.80(25°C)	3000	15.00
2SD621	Si	npn	TV_HD	2500/900	3.00	50.00($25°C)	3-15	
2SD628	Si	npn	Darl+diode,pow	100	10.00	80.00($25°C)	>1000	
2SD629	Si	npn	Darl+diode,pow	100	10.00	100.00($25°C)	>1000	
2SD688	Si	npn	Darl,pow	100	1.50	0.80($25°C)	>10	
2SD712	Si	npn	AF-s-pow	100	4.00	30.00($25°C)	55-300	8.00
2SD726	Si	npn	AF-s-pow	100/80	4.00	40.00($25°C)	35-320	10.00
2SD729	Si	npn	Darl+diode,pow	100	20.00	125.00($25°C)	>1000	
2SD781	Si	npn	s-pow,TV-HD	150/60	2.00	1.00(25°C)	150	
2SD826	Si	npn		60/20	5.00	1.00($25°C)	120-560	120.00
2SD838	Si	npn	TV-HD,s-pow	2500/900	3.00	50.00($25°C)	3-15	
2SD892A	Si	npn	Darl	60/50	0.50	0.40(25°C)	>8000	150.00
2SD1049	Si	npn	AF-s-pow	120/80	25.00	80.00($25°C)	>20	
2SD1062	Si	npn	s-pow	60/50	12.00	40.00($25°C)	>30	10.00
2SD1153	Si	npn	Darl	80750	1.50	0.90(25°C)	>40	120.00
2SD1177	Si	npn	AF-pow,TV-HD	100/60	2.00	20.00($25°C)	>40	230.00
2SD1237	Si	npn	s-pow	90/80	7.00	1.75($25°C)	>30	20.00
2SD1238	Si	npn	s-pow	90/80	12.00	80.00($25°C)	>30	20.00
2SD1639	Si	npn	AF-s-pow	100/80	2.20	10.00($25°C)	40-200	
2SD1684	Si	npn	AF/s-pow,lo-sat	120/100	1.50	10.00(25°C)	>30	120.00
2SD1685	Si	npn	AF/s-pow,lo-sat	60/20	5.00	10.00(25°C)	>95	120.00
2SD1691	Si	npn	AF-s-pow	60	5.0	20.00(25°C)	100-400	
2SD1840	Si	npn	AF/s-pow,lo-sat	110/100	15.00	100.00($25°C)	50-140	
2SD1841	Si	npn	AF/s-pow,lo-sat	110/100	25.00	120.00($25°C)	50-140	
2SD1842	Si	npn	AF/s-pow,lo-sat	110/100	40.00	150.00($25°C)	50-140	
2SD2116	Si	npn	Darl	80/50	0.70	1.00(25°C)	>40	
2SD2117	Si	npn	Darl	80/50	1.50	1.00(25°C)	>30	
2SD2213	Si	npn	Darl,AF	150/80	1.50	0.90(25°C)	>10	
2SJ165	V-MOS	p-ch	FET-enh.,	50V,0.1A,0.25W				
2SK422	V-MOS	n-ch	FET-enh.	60v,0.7A,0.9W,17/12ns				
2SK423	V_MOS	n-ch	FET-enh.	100V,0.5A,0.9W,15/20ns				
3N140	MOS	n-ch	FET-depl.,dg,FM/VHF-in	20V,Idss>5mA				
3N225	MOS	n-ch	FET-depl.,dg, UHF	25V,Idss>1mA,Up<4V				
3SK35	MOS	n-ch	FET-depl.,dg,VHF	20V,Idss>3mA,Up<4V				
3SK37	MOS	n-ch	FET-depl.,dg,VHF	20V,Idss>4mA,Up<3V				
3SK45	MOS	n-ch	FET-depl.,dg,VHF	22V,Idss>4mA,Up<3V				
3SK61	MOS	n-ch	FET-depl.,dg,VHF	20V,Idss>4mA,Up<3V				
3SK72	MOS	n-ch	FET-depl.,dg,VHF	20V,Idss>2.5mA,Up<3V				
3SK77	MOS	n-ch	FET-depl.,dg,VHF	20V,Idss>3mA,Up<2.5V				
3SK85	MOS	n-ch	FET-depl.,dg,VHF	20V,Idss>4mA,Up<3V				

SEMICONDUCTOR DIODE SPECIFICATIONS

Device	Type	Material	Peak Inverse Voltage, PIV (Volts)	Average Rectified Current Forward (Reverse) IO (A) (IR(A))	Peak Surge Current, IFSM 1 sec. @ 25ºC (A)	Average Forward Voltage, VF (Volts)
1N34	Signal	Germanium	60	8.5 m (15.0µ)		1.0
1N34A	Signal	Germanium	60	5.0 m (30.0µ)		1.0
1N67A	Signal	Germanium	100	4.0 m (5.0µ)		1.0
1N191	Signal	Germanium	90	5.0 m	1.0	
1N270	Signal	Germanium	80	0.2 (100 µ)		1.0
1N914	Fast Switch	Silicon (Si)	75	75.0 m (25.0 n)	0.5	1.0
1N1184	RFR	Si	100	35 (10 m)		1.7
1N2071	RFR	Si	600	0.75 (10.0µ)		0.6
1N3666	Signal	Germanium	80	0.2 (25.0µ)		1.0
1N4001	RFR	Si	50	1.0 (0.03 m)		1.1
1N4002	RFR	Si	100	1.0 (0.03 m)		
1N4003	RFR	Si	200	1.0 (0.03 m)		1.1
1N4004	RFR	Si	400	1.0 [0.03 m)		1.1
1N4005	RFR	Si	600	1.0 (0.03 m)		1.1
1N4006	RFR	Si	800	1.0 (0.03 m)		1.1
1N4007	RFR	Si	1000	1.0 (0.03 m)		1.1
1N4148	Signal	Si	75	10.0 m (25.0 n)		1.0
1N4149	Signal	Si	75	10.0 m (25.0 n)		1.0
1N4152	Fast Switch	Si	40	20.0 m (0.05µ)		0.8
1N4445	Signal	Si	100	0.1 (50.0 n)		1.0
1N5400	RFR	Si	50	3.0	200	
1N5401	RFR	Si	100	3.0	200	
1N5402	RFR	Si	200	3.0	200	
1N5403	RFR	Si	300	3.0	200	
1N5404	RFR	Si	400	3.0	200	
1N5405	RFR	Si	500	3.0	200	
1N5406	RFR	Si	600	3.0	200	
1N5767	Signal	Si		0.1 (1.0µ)		1.0
ECG5863	RFR	Si	600	6	150	0.9

* RFR = Rectifier, Fast Recovery

ZENER DIODES SPECIFICATIONS

Zener Voltage (Volts)	Power (Watts)							
	0.25	0.4	0.5	1.0	1.5	5.0	10.0	50.0
1.8	1N4614							
2.0	1N4615							
2.2	1N4616							
2.4	1N4617	1N4370,A	1N4370,A,1N5221,B 1N5985,B					
2.5			1N5222B					
2.6	1N702,A							
2.7	1N4618	1N4371,A	1N4371,A,1N5223,B 1N5839, 1N5986					
2.8			1N5224B					
3.0	1N4619	1N4372,A	1N4372,1N5225,B 1N5987					
3.3	1N4620	1N746,A 1N764 A 1N5518	1N746A 1N5226,B 1N5988	1N3821 1N4728,A	1N5913	1N5333,B		
3.6	1N4621	1N747,A 1N5519	1N747A 1N5227,B,1N5989	1N3822 1N4729,A	1N5914	1N5334,B		
3.9	1N4622	1N748,A 1N5520	1N748A,1N5228,B 1N5844, 1N5990	1N3823 1N4730,A	1N5915	1N5335,B	1N3993A	1N4549,B 1N4557,B
4.1	1N704,A							
4.3	1N4623	1N749,A 1N5521	1N749,A 1N5229,B 1N5845,1N5991	1N3824 1N4731,A	1N5916	1N5336,B	1N3994,A	1N4550,B 1N4558,B
4.7	1N4624	1N750,A 1N5522	1N750A,1N5230,B 1N5846, 1N5992	1N3825 1N4732,A	1N5917	1N5337,B	1N3995,A	1N4551,B 1N4559,B
5.1	1N4625 1N4689	1N751 A 1N5523	1N751A, 1N5231,B 1N5847,1N5993	1N3826 1N4733	1N5918	1N5338,B 1N4560,B	1N3996,A	1N4552,B
5.6	1N708A 1N4626	1N752,A 1N5524	1N752,A,1N5232,B 1N5848, 1N5994	1N3827 1N4734,A	1N5919	1N5339,B 1N4561,B	1N3997,A	1N4553,B
5.8	1N706A	1N762						
6.0				1N5233B 1N5849			1N5340,B	
6.2	1N709,1N4627 MZ605,MZ610 MZ620,MZ640	1N753,A 1N821,3,5, 7,9; A	1N753,A 1N5234,B, 1N5850 1N5995	1N3828,A 1N4735,A	1N5920	1N5341,B 1N4562,B	1N3998,A	1N4554,B
6.4	1N4565-84,A							
6.8	1N4099	1N754,A 1N957,B 1N5526	1N754,A 1N757,B 1N5235,B 1N5851 1N5996	1N3016,B 1N3829 1N4736,A	1N3785 1N5921	1N5342,B	1N2970,B 1N3999,A	1N2804B 1N3305B 1N4555, 1N4563
7.5	1N4100	1N755,A 1N958,B 1N5527	1N755A,1N958,B 1N5236,B, 1N5862 1N5997	1N3017,A,B 1N3830 1N4737,A	1N3786 1N5922	1N5343,B 1N4000,A 1N4556,	1N2971,B 1N3306,B	1N2805,B 1N4564
8.0	1N707A							
8.2	1N712A 1N4101	1N756,A 1N959,B 1N5528	1N756,A 1N959,B,1N5237,B 1N5853,1N5998	1N3018,B 1N4738,A	1N3787 1N5923	1N5344,B	1N2972,B	1N2806,B 1N3307,B
8.4		1N3154-57,A 1N3155-57	1N3154,A					
8.5	1N4775-84,A		1N5238,B,1N5854					
8.7	1N4102					1N5345,B		
8.8		1N 764						
9.0		1N764A	1N935-9;A,B					

ZENER DIODES SPECIFICATIONS

Zener Voltage (Volts)	Power (Watts)							
	0.25	0.4	0.5	1.0	1.5	5.0	10.0	50.0
9.1	1N4103	1N757,A 1N960,B 1N5529	1N757,A, 1N960,B 1N5239,B, 1N5855 1N5999	1N3019,B 1N4739,A	1N3788 1N5924	1N5346,B	1N2973,B	1N2807,B 1N3308,B
10.0	1N4104	1N758,A 1N961,B 1N5530,B	1N758,A, 1N961,B 1N5240,B, 1N5856 1N6000	1N3020,B 1N4740	1N3789 1N5925	1N5347,B	1N2974,B	1N2808,B 1N3309,A,B
11.0	1N715,A 1N4105	1N962,B 1N5531	1N962,B,1N5241,B 1N5857, 1N6001	1N3021,B 1N4741,A	1N3790 1N5926	1N5348,B	1N2975,B	1N2809,B 1N3310,B
11.7	1N716,A 1N4106		1N941,A,B					
12.0		1N759,A 1N963,B 1N5532	1N759,A ,1N963,B 1N5242,B, 1N5858 1N6002	1N3022,B 1N4742,A	1N3791 1N5927	1N5349,B	1N2976,B	1N2810,B 1N3311,B
13.0	1N4107	1N964,B 1N5533	1N964,B,1N5243,B 1N5859,1N6003	1N3023,B 1N4743,A	1N3792 1N5928	1N5350,B	1N2977,B	1N2811,B 1N3312,B
14.0	1N4108	1N5534	1N5244B, 1N5860			1N5351,B	1N2978,B	1N2812,B 1N3313,B
15.0	1N4109	1N965,B 1N5535	1N965,B,1N5245,B 1N5861,1N6004	1N3024,B 1N4744A	1N3793 1N5929	1N5352,B	1N2979,A,B	1N2813,A,B 1N3314,B
16.0	1N4110	1N966,B 1N553,B	1N966,B,1N5246,B 1N5862, 1N6005	1N3025,B 1N4745,A	1N3794 1N5930	1N5353,B	1N2980,B	1N2814,B 1N3315,B
17.0	1N4111	1N5537	1N5247,B 1N5863			1N5354,B	1N2981B	1N2815,B 1N3316,B
18.0	1N4112	1N967,B 1N5538	1N967,B 1N5248,B 1N5864, 1N6006	1N3026,B 1N4746,A	1N3795 1N5931	1N5355,B	1N2982,B	1N2816,B 1N3917,B
19.0	1N4113	1N5539	1N5249,B 1N5865			1N5356,B	1N2983,B	1N2817,B 1N3318,B
20.0	1N4114	1N968,B 1N5540	1N968,B,1N5250,B 1N5866, 1N6007	1N3027,B 1N4747,A	1N3796 1N5932,A,B	1N5357,B	1N2984,B	1N2818,B 1N3319,B
22.0	1N4115	1N959,B 1N5541	1N969,B,1N5241,B 1N5867, 1N6008	1N3028,B 1N4748,A	1N3797 1N5933	1N5358,B	1N2985,B	1N2819,B 1N3320,A,B
24.0	1N4116	1N5542 1N9701B	1N970,B,1N5252,B 1N586,1N6009	1N3029,B 1N4749,A	1N3798 1N5934	1N5359,B	1N2986,B	1N2820,B 1N3321,B
25.0	1N4117	1N5543	1N5253,B 1N5869			1N5360,B	1N2987B	1N2821,B 1N3322,B
27.0	1N4118	1N971,B	1N971,1N5254,B 1N5870,1N6010	1N3030,B 1N4750,A	1N3799 1N5935	1N5361,B	1N2988,B	1N2822B 1N3323,B
28.0	1N4119	1N5544	1N5255,B,1N5871			1N5362,B		
30.0	1N4120	1N972,B 1N5546	1N972,B,1N5256,B 1N5872,1N6011	1N3031,B 1N4751,A	1N3800 1N5936	1N5363,B	1N2989,B	1N2823,B 1N3324,B
33.0	1N4121	1N973,B 1N5546	1N973,B,1N5257,B 1N5873,1N6012	1N3032,B 1N4752,A	1N3801 1N5937	1N5364,B	1N2990,A,B	1N2824,B 1N3325,B
36.0	1N4122	1N974,B	1N974,B,1N5258,B 1N5874,1N6013	1N3033,B 1N4753,A	1N3802 1N5938	1N5365,B	1N2991,B	1N2825,B 1N3326,B
39.0	1N4123	1N975,B	1N975,B, 1N5259,B 1N5875 ,1N6014	1N3034,B 1N4754,A	1N3803 1N5939	1N5366,B	1N2992,B	1N2826,B 1N3327,B
43.0	1N4124	1N976,B	1N976,B,1N5260,B 1N5876,1N6015	1N3035,B 1N4755,A	1N3804 1N5940	1N5367,B	1N2993,A,B	1N2827,B 1N3328,B
45.0			1N2994B	1N2828B 1N3329B				

POWER FETs

Device No.	Type	Max. Diss. (W)	Max. VDS (Volts)	Max ID (A)*	Gfs mmhos (typ.)	Input Ciss (pF)	Output Coss (pF)	Approx. Upper Freq. (MHz)	Case	Pack-Type Mnfr.	General applications age/
DV1202S	N-Chan.	10	50	0.5	100k	14	20	500	.380 SOE	1/S	RF power amp., oscillator
DV1202W	N-Chan.	10	50	0.5	100k	14	20	500	C-220	5/S	RF power amp., oscillator
DV1205S	N-Chan.	20	50	1	200k	26	38	500	.380 SOE	1/S	RF power amp., oscillator
DV1205W	N-Chan.	20	50	1	200k	26	98	500	C-220	5/S	RF power amp., oscillator
2SK133	N-Chan.	100	120	7	1M	600	350	1	TO-3	6/H	AF pwr. amp., switch (complem to 25J48)
2SK134	N-Chan.	100	140	7	1M	600	350	1	TO-3	6/H	AF pwr. amp., switch (complem to 25J49)
2SK135	N-Chan.	100	160	7	1M	600	350	1	TO-3	6/H	AF pwr. amp., switch (complem to 25J50)
2SJ48	P-Chan.	100	120	7	1M	900	400	1	TO-3	6/H	AF pwr. amp., switch (complem to 2SK133)
2SJ49	P-Chan.	100	140	7	1M	900	400	1	TO-3	6/H	AF pwr. amp., switch (complem to 2SK134)
2SJ50	P-chan.	100	160	7	1M	900	400	1	TO-3	6/H	AF pwr. amp., switch (complem to 2SK135)
VMP4	N-Chan.	25	60	2	170K	32	4.8	200	.380 SOE	1/S	VHF pwr. amp., rcvr front end (rf amp., mixer).
VN10KM	N-Chan.	1	60	0.5	100K	48	16	-	TO-92	2/S	High-speed line driver, relay driver, LED stroke driver
VN64GA	N-Chan.	80	60	12.5	150K	700	325	30	TO-3	3/S	Linear amp., power-supply switch, motor control
VN66AF	N-Chan.	15	60	2	150K	50	50	-	TO-202	4/S	High-speed switch, HF linear amp., audio amp. line driver.
VN66AK	N-Chan.	8.3	60	2	250K	93	6	100	TO-39	7/S	RF pwr. amp.,high-current analog switching
VN67AJ	N-Chan.	25	60	2	250K	33	7	100	TO-3	3/S	RF pwr. amp.,high-current switching
VN89AA	N-Chan.	25	80	2	250K	50	10	100	TO-3	3/S	High-speed switching,HF linear amps., line drivers.
IRF100	N-Chan.	125	80	16	300K	900	25	-	TO-3	3/S	High-speed switching,audio inverters.
IRF101	N-Chan.	125	60	16	300K	900	25	-	TO-3	3/S	Same as IRF100

Legend: * 25°C (case) S = M/A-COM H = Hitachi IR = International Rectifier. Mnfr = Manufacturer

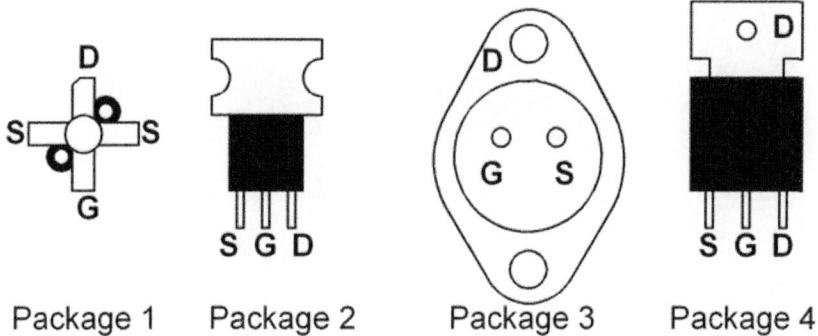

Package 1   Package 2   Package 3   Package 4

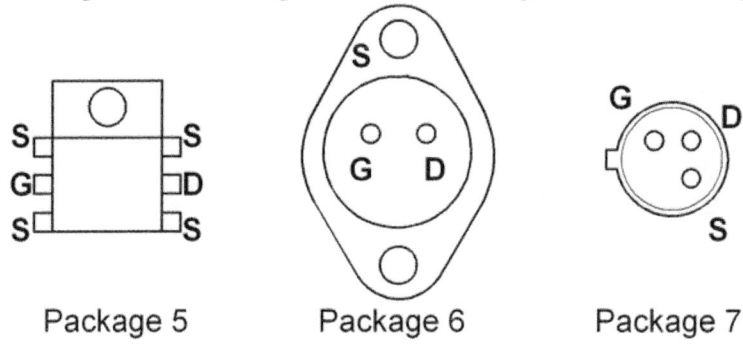

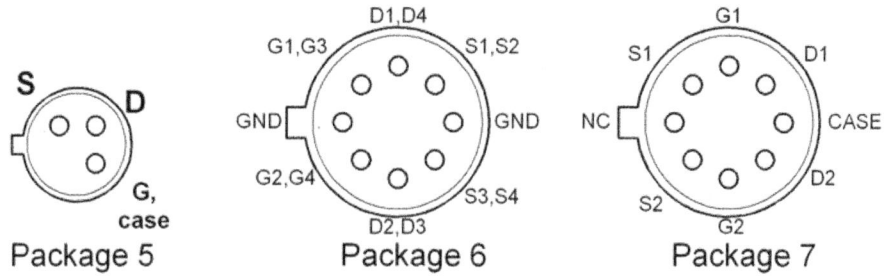

Package 5   Package 6   Package 7

Package Information for Power FETs

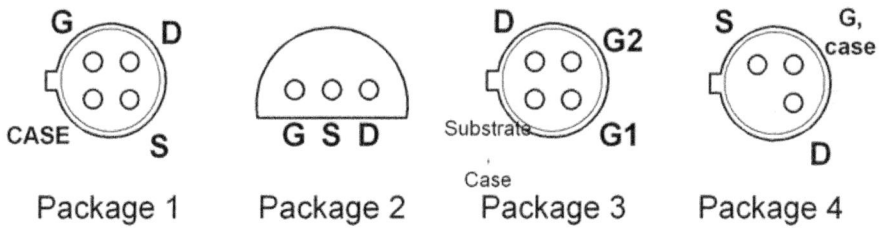

Package 1   Package 2   Package 3   Package 4

Package 5   Package 6   Package 7

Package Information for Small Signal FETs

SMALL-SIGNAL FETs

Device No.	Type	Max. Diss. (mW)	Max. VDS (Volts)	Max ID	Min Gfs (mA)*	Input C (mS)	VGS(off) (pF)	Upper Freq. (volts)(MHz)	Noise Figure (MHz)	Case Type (typ)	/Mnfr.	General applications
2N4416	N-JFET	300	30	-15	4.5K	4	-6	450	400 MHz 4 dB	TO-72	1/S,M	VHF/UHF/RF amp.mix., osc.
2N5484	N-JFET	310	25	30	2.5K	5	-3	200	200 MHz 4 dB	TO-92	2/M	VHF/UHFamp.,mix., osc.
2N5485	N-JFET	310	25	30	3.5K	5	-4	400	400 MHz 4 dB	TO-92	2/S	VHF/UHF/RF amp.mix., osc.
3N200	N-Dual-Gate MOSFET	330	20	50	10K	4-8.5	-6	500	400 MHz 4.5 dB	TO-72	3/R	VHF/UHF/RF amp.mix., osc.
3N202	N-Dual-Gate MOSFET	360	25	50	8K	6	-5	200	200 MHz 4.5 dB	TO-72	3/S	VHF amp., mixer
MPF102	N-JFET	310	25	20	2K	4.5	-8	200	200 MHz	TO-92	2/N,M	HF/VHF amp.,mix., osc.,
MPF106/ 2N5484	N-JFET	310	25	30	2.5K	5	-6	400	200 MHz 4 dB	TO-92	2/N,M	HF/VHF/UHF amp.,mix.,osc.
40673	N-Dual-Gate MOSFET	330	20	50	12K	6	-4	400	200 MHz 6 dB	TO-72	3/R	HF/VHF/UHF amp. mix., osc.
U300	P-JFET	300	-40	20	8K	-50	+10	-	400 MHz	TO-18	4/S	General Purpose amp.
U304	P-JFET	350	-30	-50	10K	27	+10	-	-	TO-18	4/S	analog switch, chopper
U310	N-JFET	500	30	60	10K	2.5	-6	450	450 MHz 3.2 dB	TO-52	5/S	common-gate VHF/UHF amp.,osc., mixer
U350	N-JFET Quad	300 1W	25	60	9K	5	-6	100	100 MHz 7 dB	TO-99	6/S	matched JFET doubly bal. mixer
U431	N-JFET Dual	300	25	30	10K	5	-6	100	100 MHz -	TO-99	7/S	matched JFET cascade amp., balanced mixer

* 25°C

S = Siliconix Inc.   R = RCA   N = National Semiconductor   M = Motorola

Three-Terminal Voltage Regulators

* Listed numerically by device

Device	Description	Voltage	Current (Amps)	Package
317	Adj. Pos	+1.2 to +37	0.5	TO-205
317	Adj. Pos	+1.2 to +37	1.5	TO-204,TO-220
317L	Low Current Adj. Pos	+1.2 to +37	0.1	TO-205,TO-92
317M	Med Current Adj. Pos	+1.2 to +37	0.5	TO-220
350	High Current Adj. Pos	+1.2 to +33	3.0	TO-204,TO-220
337	Adj. Neg	-1.2 to -37	0.5	TO-205
337	Adj. Neg	-1.2 to -37	1.5	TO-204,TO-220
337M	Med Current Adj. Neg	-1.2 to -37	0.5	TO-220
309		+5	0.2	TO-205
309		+5	1.0	TO-204
323		+5	9.0	TO-204,TO-220
140-XX	Fixed Pos	Note #	1.0	TO-204,TO-220
340-XX			1.0	TO-204,TO-220
78XX			1.0	TO-204,TO-220
78LXX			0.1	TO-205,TO-92
78MXX			0.5	TO-220
78TXX			3.0	TO-204
79XX	Fixed Neg	Note #	1 .0	TO-204,TO-220
79LXX			0.1	TO-205,TO-92
79MXX			0.5	TO-220

Legend:

Adj.	= Adjustable
Med	= Medium
Neg	= Negative
Pos	= Positive

Note # - XX indicates the regulated voltage; which may be anywhere from 1.2 volts to 35 volts. For example a 7808 is a positive 8-volt regulator, and a 7912 is a negative 12-volt regulator.

The regulator package may be denoted by an additional suffix, according to the following:

Package	Suffix
TO-204 (TO-3)	K
TO-220	T
TO-205 (TO-39)	H,G
TO-92	P,Z

Example:

A 7815K is a positive 15-volt regulator in a TO-204 package. An LM340T-8 is a positive 8-volt regulator in a TO-220 package. In addition, different manufacturers use different prefixes. An LM7812 is equivalent to a μA 7812 or MC7812.

P,Z SUFFIX TO-92 PACKAGE

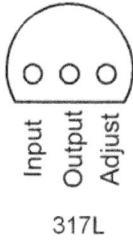

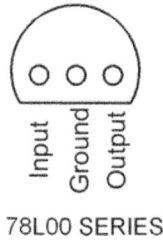

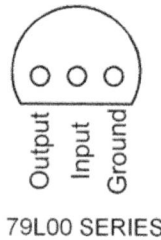

317L — Input, Output, Adjust

78L00 SERIES — Input, Ground, Output

79L00 SERIES — Output, Input, Ground

T SUFFIX TO-220 PACKAGE

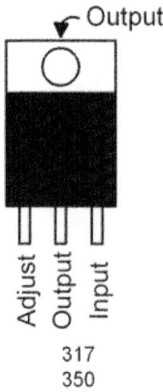

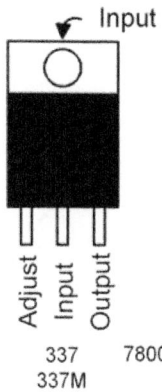

Output — Adjust, Output, Input
317
350

Input — Adjust, Input, Output
337
337M

Ground — Input, Ground, Output
7800 SERIES; 78T00 SERIES
87M00 SERIES
140T-XX; 340T-XX

Input — Ground, Input, Output
7900 SERIES
79M00 SERIES

H,G SUFFIX TO-205 PACKAGE

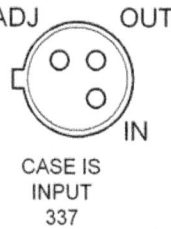

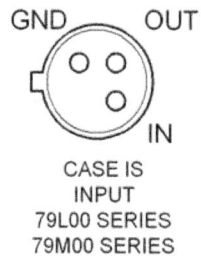

IN — ADJ — OUT
CASE IS OUTPUT
317
317L

ADJ — OUT — IN
CASE IS INPUT
337

IN — OUT — GND
CASE IS GROUND
78L00 SERIES
78M00 SERIES

GND — OUT — IN
CASE IS INPUT
79L00 SERIES
79M00 SERIES

K SUFFIX METAL TO-204 PACKAGE

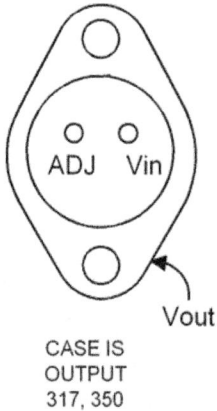

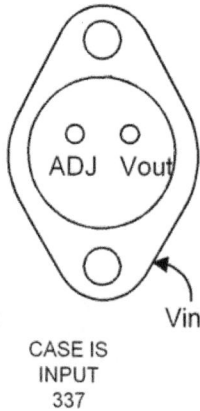

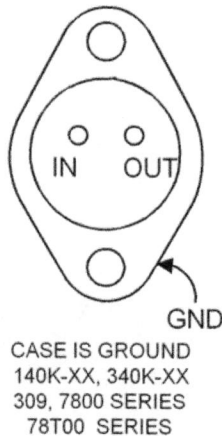

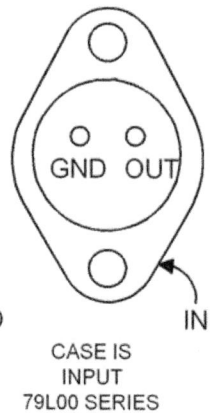

ADJ Vin — Vout
CASE IS OUTPUT
317, 350

ADJ Vout — Vin
CASE IS INPUT
337

IN OUT — GND
CASE IS GROUND
140K-XX, 340K-XX
309, 7800 SERIES
78T00 SERIES

GND OUT — IN
CASE IS INPUT
79L00 SERIES

PRINTED CIRCUIT BOARD LAYOUTS

All printed circuit board layouts in this collection are once again printed in the following pages. You can either cut out or photocopy these pages to make a separate file for quick reference.

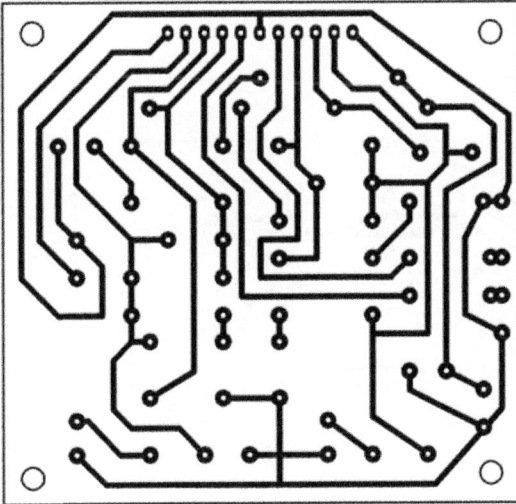

page 18 Car Stereo Booster

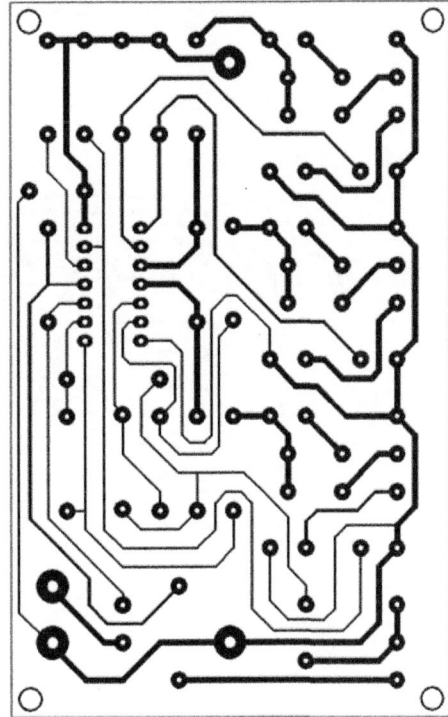

page 15 Audio Mixer

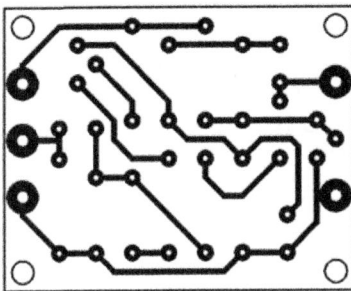

page 11 Audio Compressor

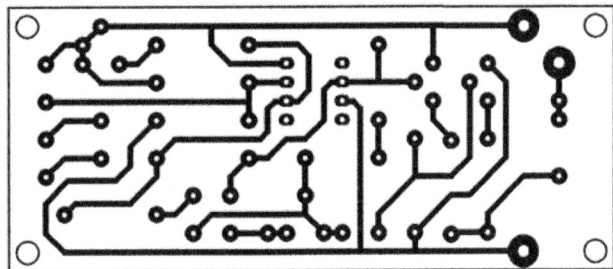

page 13 Universal Amplifier

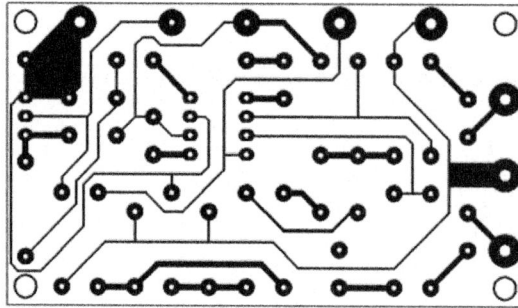

page 22   Cardiophone

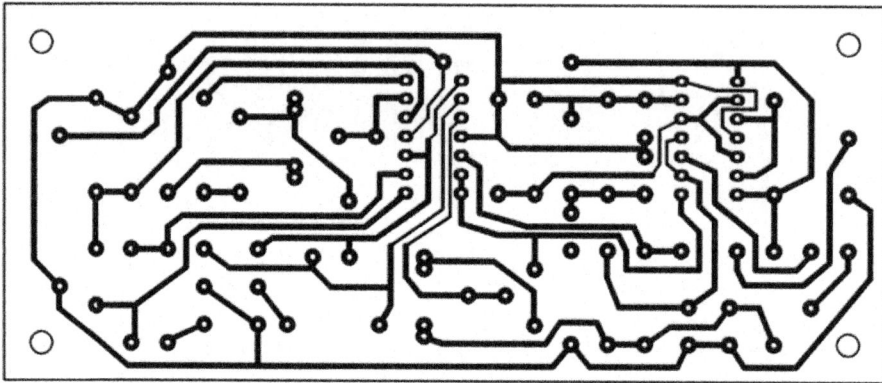

page 26   Audio Squelch

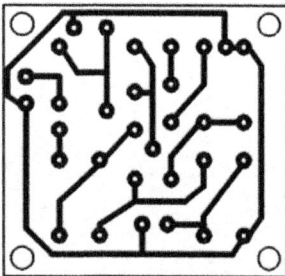

page 29   Low Noise Preamp

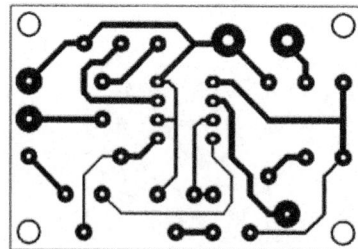

page 30   Mini Audio Amp

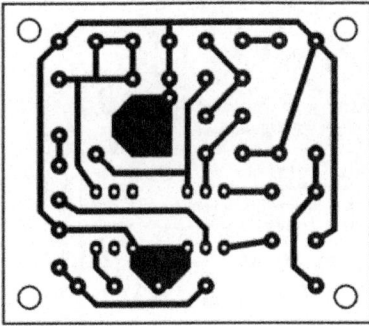

page 31 6.5 Watts Amplifier

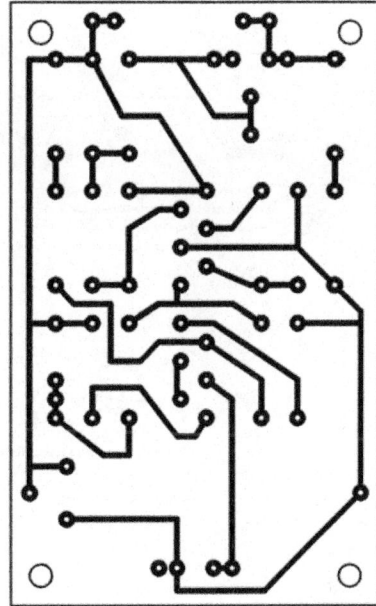

page 37 Speaker Peak Indicator

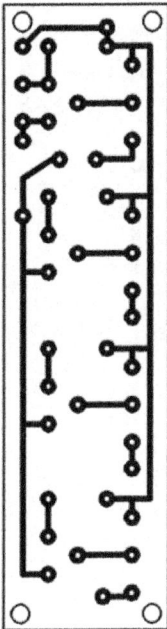

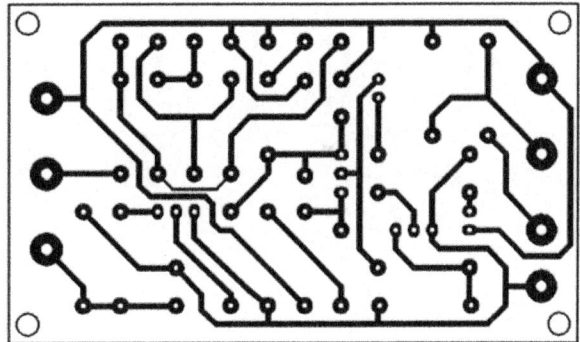

page 39 Automatic Volume Control

page 34 FET Audio Mixer

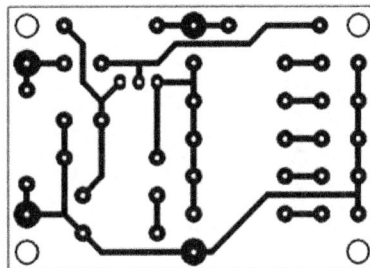

page 41 Audio Mixer

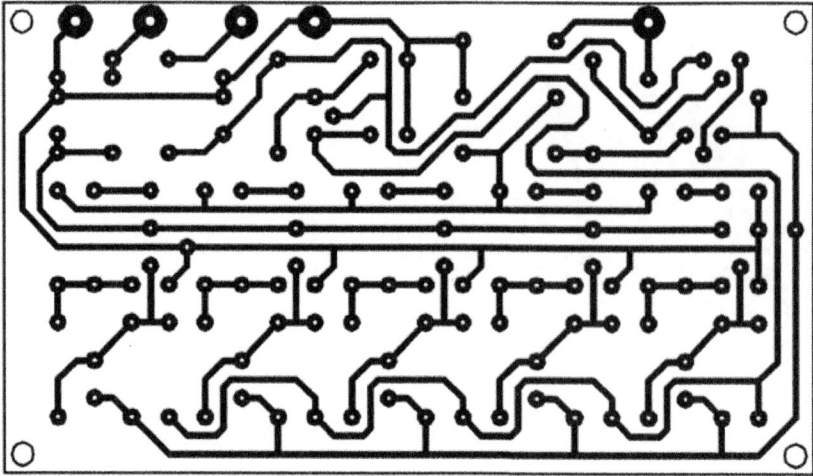

page 42   Very Low Noise Mic Amp

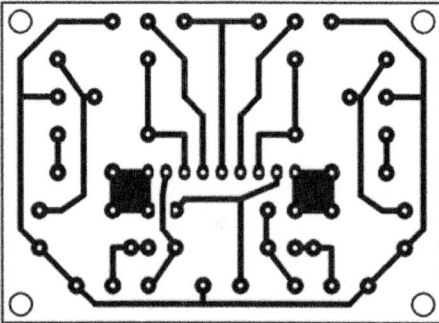

page 44   Cassette Preamp

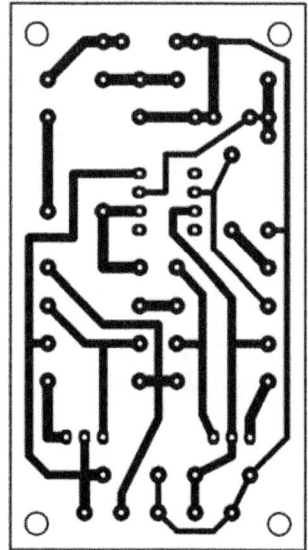

page 60   HIFI Headphone Amp

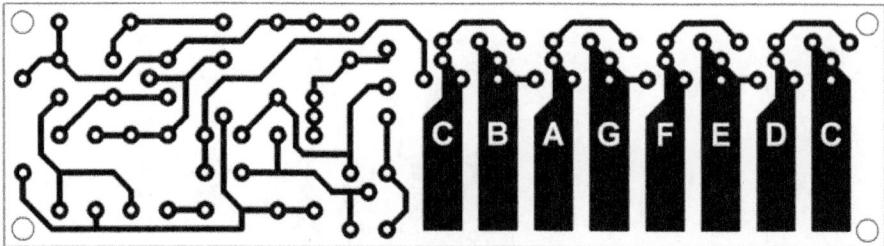

page 54   Electronic Organ

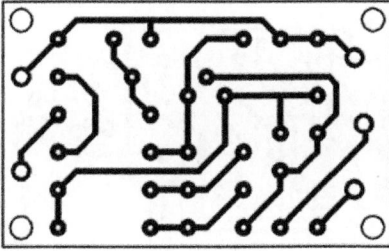

page 62   AF Generator

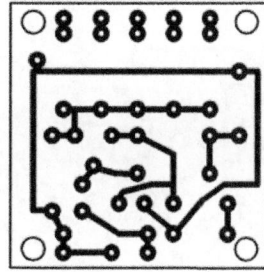

page 66   Audio Mixer

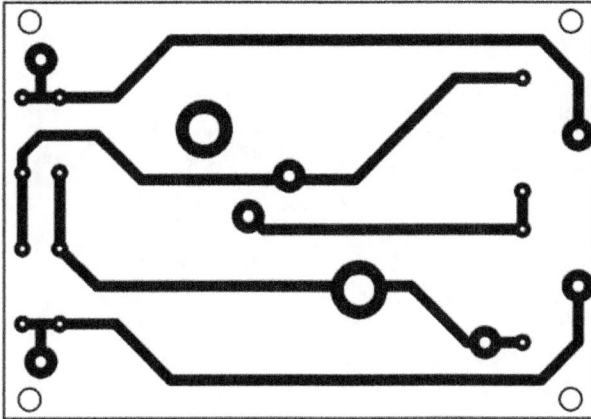

page 64   Tweeter Guardian

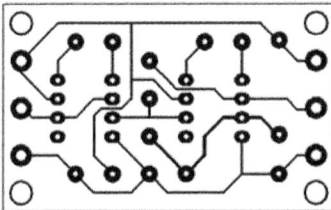

page 67   Touch Volume Control

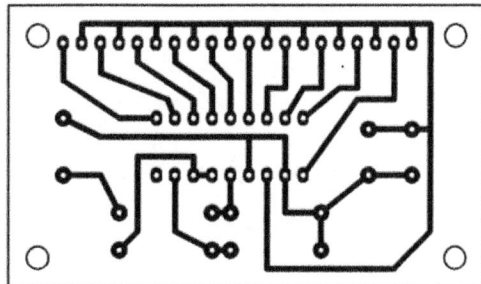

page 69   Audio Wattmeter

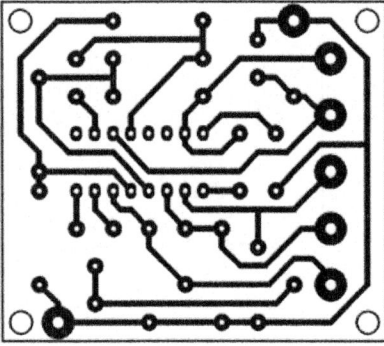

page 70 Low Noise Preamp

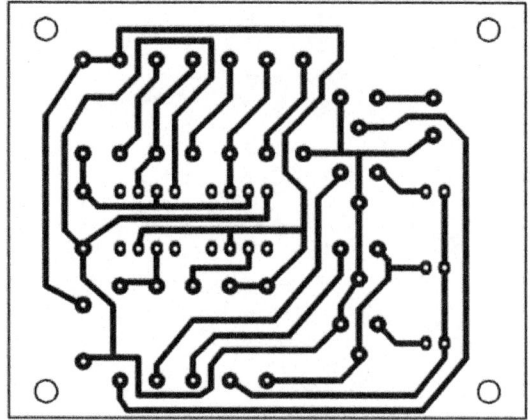

page 72 Heatsink Thermometer

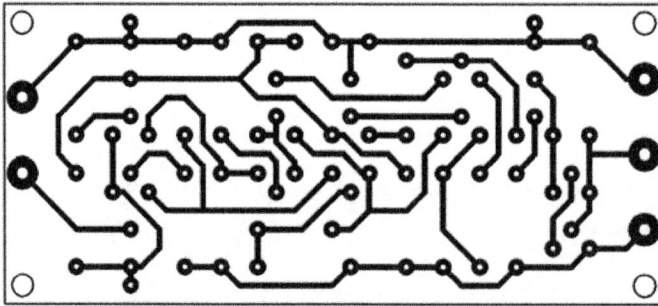

page 76 Audio Peak Meter

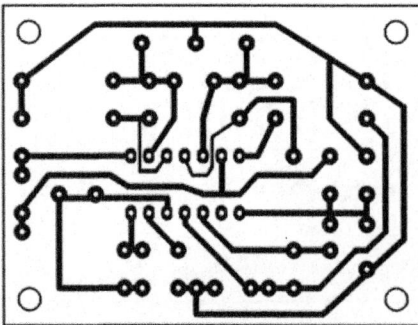

page 80 Microphone Preamp

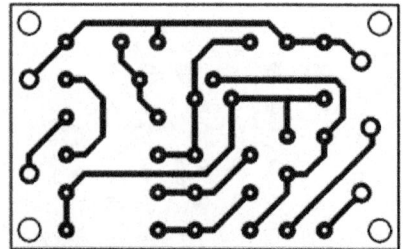

page 85 AF Generator

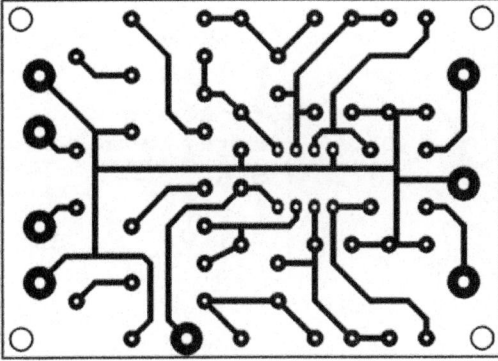

page 86 Stereo Audio Mixer

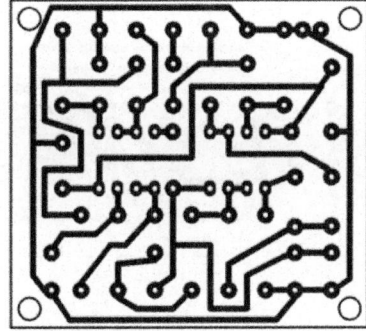

page 89 Audio Filter

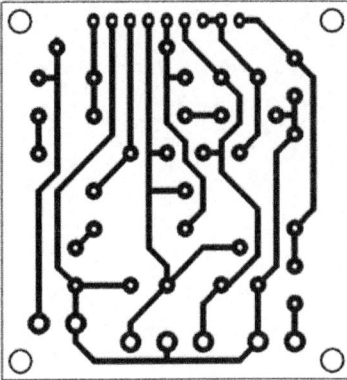

page 90 1-Chip 40 Watt Amp

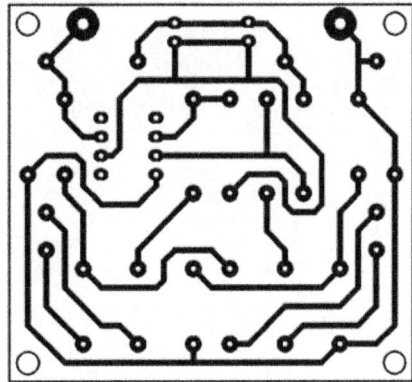

page 93
Balance Indicator

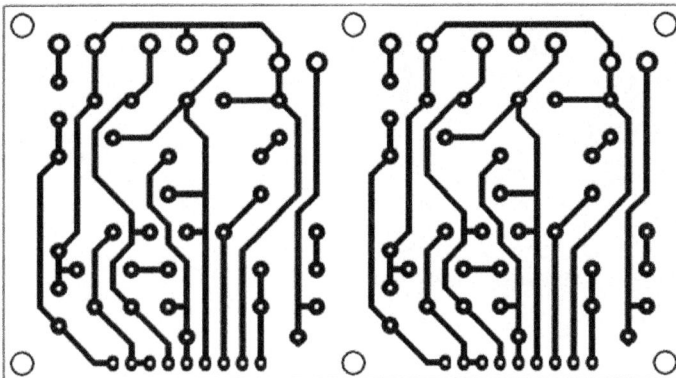

page 90 1-Chip 40Watt Amp (stereo version)

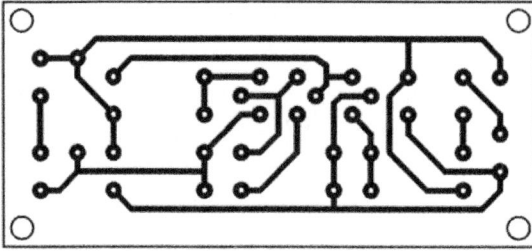

page 95 Video Amplifier

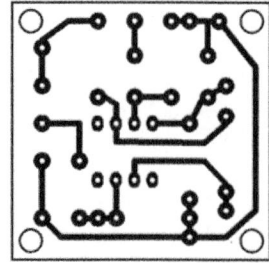

page 96
E-Guitar Preamp

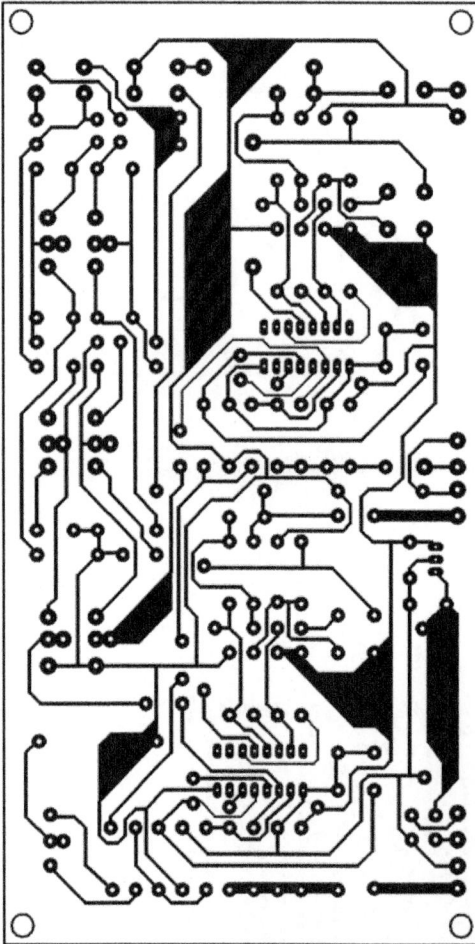

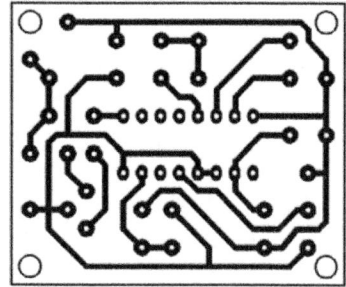

page 108
Music in Chip

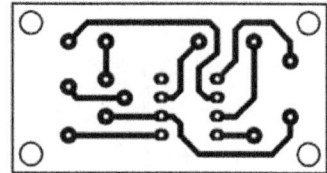

page 121 Multisound Siren

page 98 HIFI Stereo Preamp

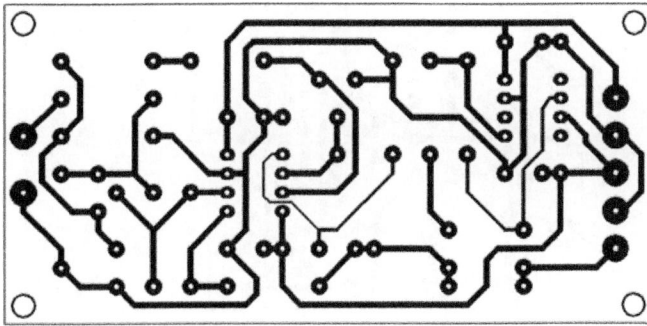

page 106   Auto Volume Control

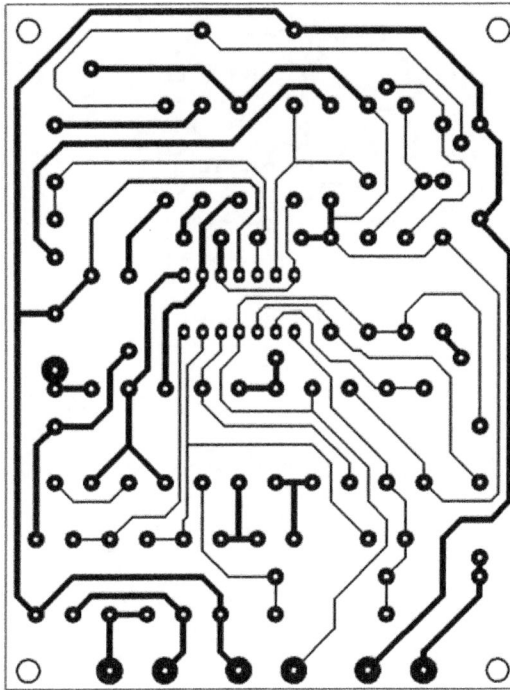

page 110   Audio Tester

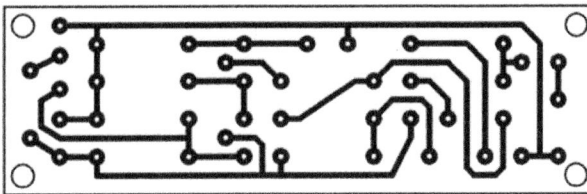

page 125   Telephone Amplifier

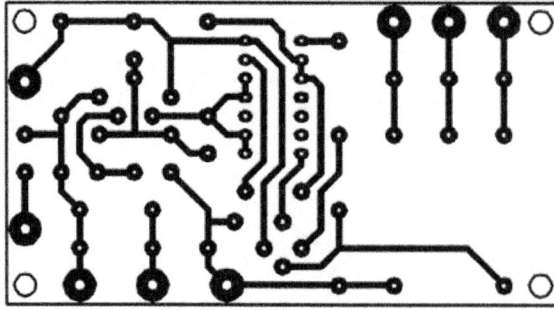

page 130   VOX Switch

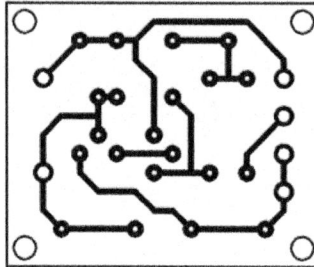

page 134   Siren

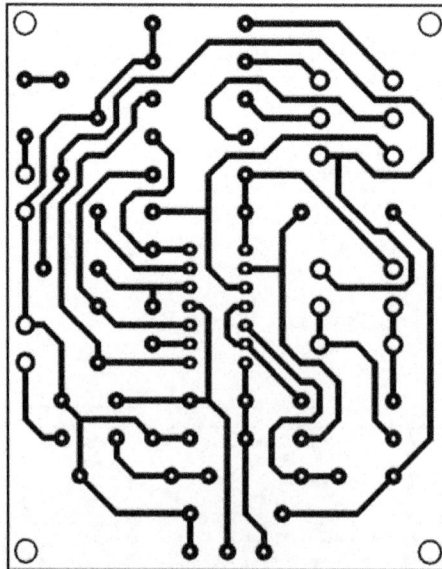

page 119   DX Audio Filter

This page intentionally left blank.

Index

Index

Index

Index

U

Ultrasound 112
universal audio 31
universal preamp 51

V

vibrato 54
vibrato generator 137
video amplifier 95
voice operated 130
Voice regulation 11
voltage controlled resistor 123
Volume Control 106, 139
VOX 130

W

wienbridge 111

X

XR2206 122

Z

zener voltage 72

Notes

Notes

Notes

Notes

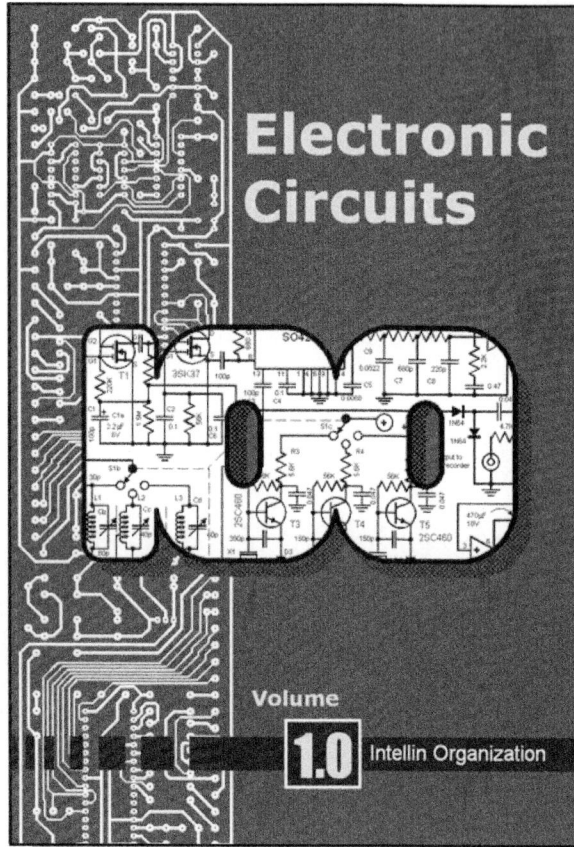

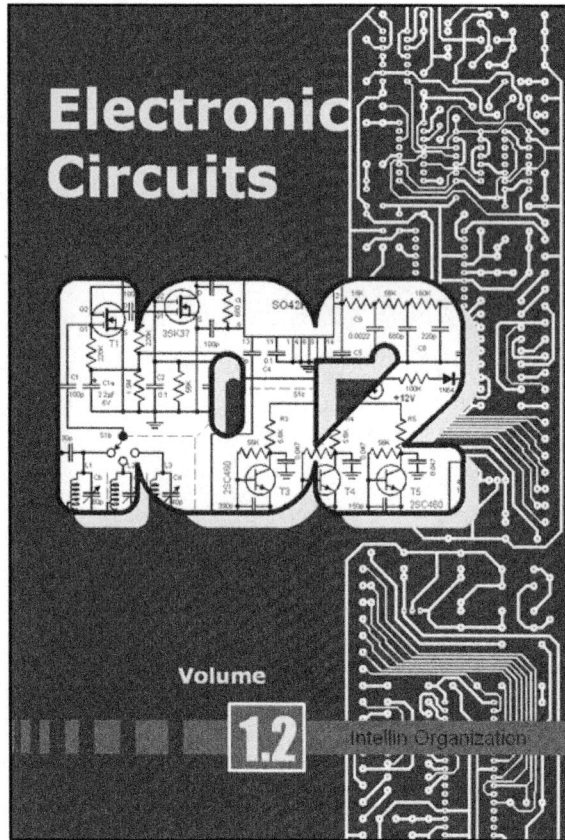

ONE HUNDRED and THREE circuits with
ready to use pcb design layouts!

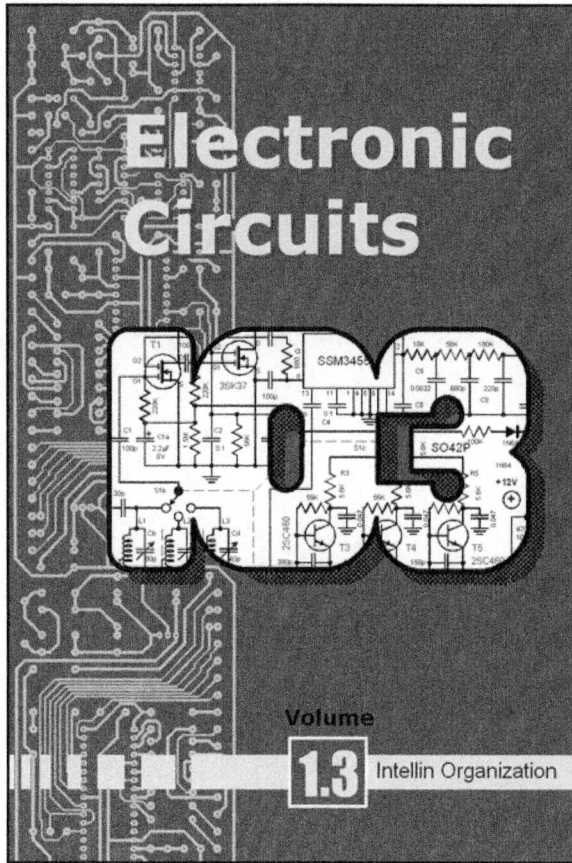

Get your copy now from
amazon.com!

This authoritative and well-researched book is the only one available that will give you all of the most important and reliable on VHF antenna construction techniques.

This unique book offers a superb collection of detailed, easy-to-follow, fully illustrated, and tested designs, covering such types of antennas as:

Omnidirectional antennas

Gain-omni antennas

Gain-directed beams

Portable antennas

Yagi antennas

Stacked arrays

Stacked collinears

Wideband-omni antennas

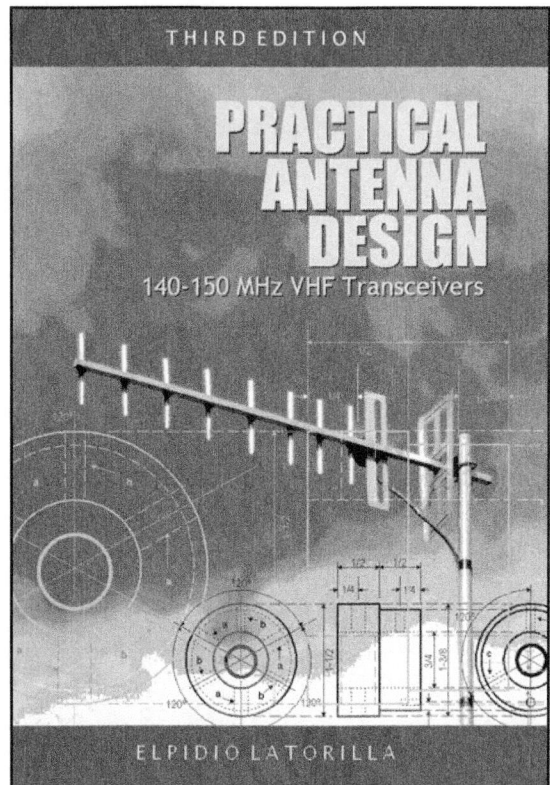

THIRD EDITION

PRACTICAL ANTENNA DESIGN

140-150 MHz VHF Transceivers

ELPIDIO LATORILLA

Get your copy now from amazon.com!

Printed in Great Britain
by Amazon